A TAXONOMIC REVISION OF
NEARCTIC *ENDASYS* FOERSTER 1868
(HYMENOPTERA: ICHNEUMONIDAE, GELINAE)

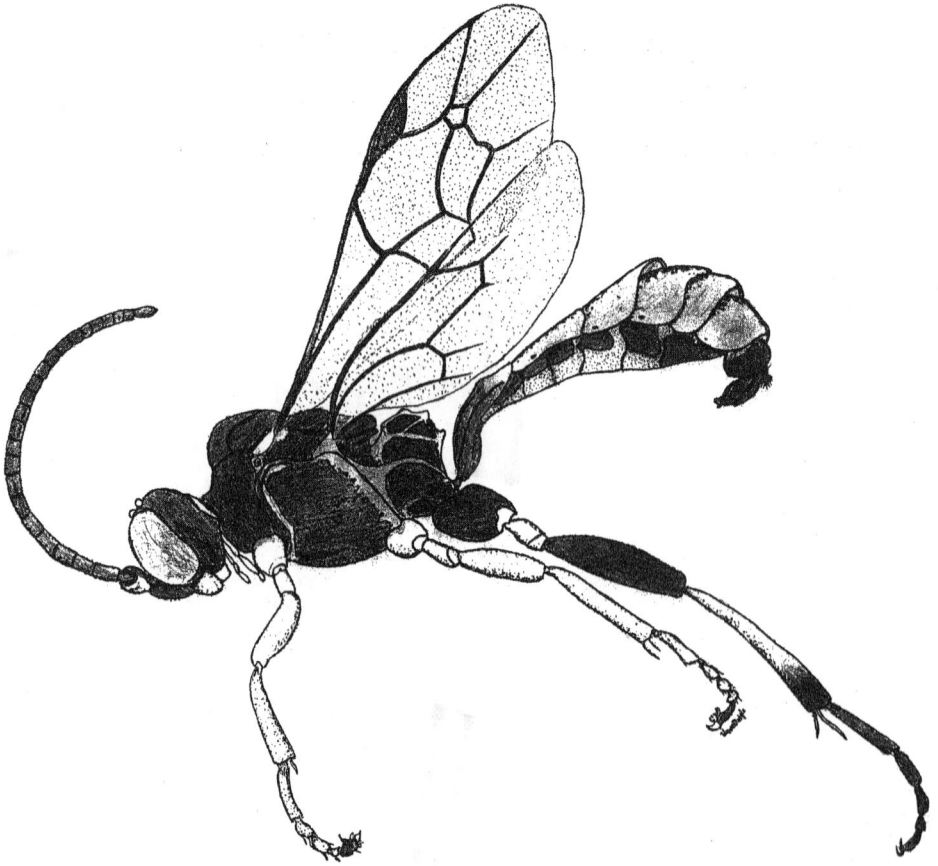

Endasys pubescens (Prov.)
Habitus

A Taxonomic Revision of Nearctic *Endasys* Foerster 1868 (Hymenoptera: Ichneumonidae, Gelinae)

John C. Luhman

UNIVERSITY OF CALIFORNIA PRESS

Berkeley • Los Angeles • Oxford

UNIVERSITY OF CALIFORNIA PUBLICATIONS IN ENTOMOLOGY

Volume 109
Issue Date: December 1990

UNIVERSITY OF CALIFORNIA PRESS
BERKELEY AND LOS ANGELES, CALIFORNIA

UNIVERSITY OF CALIFORNIA PRESS, LTD.
OXFORD, ENGLAND

Library of Congress Cataloging-in-Publication Data

Luhman, John C.
    A Taxonomic revision of Nearctic Endasys Foerster 1868
(Hymenoptera, Ichneumonidae, Gelinae) / John C. Luhman
          p.   cm. — (University of California Publications in
entomology; vol. 109)
    Includes bibliographical references.
    ISBN 0-520-09757-2 (alk. paper)
    1. Endasys—Classification. I. Title. II. Series.
QL568.I2L84   1990
595.79—dc20   89-20636
                                                    CIP

Contents

Acknowledgments

This study is based on Nearctic specimens borrowed from museums in Canada and the United States. Some type material was loaned by museums in Denmark and Sweden. Palearctic specimens were studied at the Institute of Zoology, Polish Academy of Sciences, Warsaw, as well as in collections from the Soviet Union (ZIL) and North America. Listed below are the institutions and individuals that loaned material, in alphabetical order according to acronyms.

AEI   Henry and Marjorie Townes Collection, American Entomological Institute, Gainesville, Fla. (H. and M. Townes)

AMNH   American Museum of Natural History, New York, N.Y. (M. Favreau)

ANSP   Academy of Natural Sciences, Philadelphia, Pa. (D. Azuma)

CAFA   California State Collection of Arthropods, California Department of Food and Agriculture, Sacramento (M. Wasbauer)

CAS   California Academy of Sciences, San Francisco (W. J. Pulawski)

CFS   Canadian Forestry Service, Ste.-Foy, Québec, Canada (J.-P. Laplante)

CNC   Canadian National Insect Collection, Ottawa, Ontario (J. R. Barron)

CU   Cornell University, Ithaca, N.Y. (J. K. Liebherr)

Dasch   Clement E. Dasch Collection, Muskingum College, New Concord, Ohio (C. E. Dasch)

FLDA   Florida State Collection of Arthropods, Florida Department of Agriculture, Gainesville (J. Wiley)

INF   Institute of Northern Forestry, Fairbanks, Alaska (R. Werner)

MCZ   Museum of Comparative Zoology, Harvard University, Cambridge, Mass. (S. R. Shaw)

MPM   Milwaukee Public Museum, Milwaukee, Wis. (G. R. Noonan)

NHM   Natural History Museum of Los Angeles County, Calif. (R. Snelling)

NRS   Naturhistorika Riksmuseet, Stockholm, Sweden (T. Kronestedt)

OHSU   Ohio State University, Columbus (C. A. Triplehorn)

OSU   Oregon State University, Corvallis (K. A. Phillips)

UAC   University of Alberta, Edmonton, Canada (D. Shpeley)

UAM   University of Alabama, Monticello (L. C. Thompson)

UCB   University of California, Berkeley (L. E. Caltagirone)
UCD   University of California, Davis (R. M. Bohart)
UCR   University of California, Riverside (J. Hall)
UGA   University of Georgia, Athens (C. L. Smith)
UKS   University of Kansas, Snow Museum, Lawrence (E. Cook and R. W. Brooks)
UL   Université Laval, Québec, Québec, Canada (J.-M. Perron)
UM   University of Minnesota, St. Paul (P. Clausen)
USNM   United States National Museum, Washington, D.C. (T. P. Nuhn)
UZIL   Universitetets Zoologiska Institutionen, Lund, Sweden (Thomson Collection)
Yu   Dicki Yu Collection, (currently) Agriculture Canada Research Station
        Lethbridge, Alberta (D. Yu)
ZIL   Zoological Institute, Academy of Sciences, Leningrad, U.S.S.R. (D. R.
        Kasparyan)
ZMC   Zoologisk Museum, Copenhagen, Denmark (B. Petersen)

In addition I wish to thank Dr. J. McNeil (UL) and his students for their hospitality while I was in Québec to study the Provancher types. Special thanks go to Dr. Scott Shaw (MCZ) who found the missing syntype of *E. paludicola* (Brues).

The habitus drawing of *E. pubescens* is modified from an original drawing by Mrs. Urve Daigle (UM).

Note: Author's present address: Dr. John C. Luhman, Department of Entomology, Pennsylvania State University, State College, Pennsylvania 16802.

Abstract

Endasys Foerster (Gelini, Endaseina = Phygadeuontinae, Endaseini) is revised for the Nearctic. Seventy-seven (77) species are recognized in 6 species groups. Two species are resurrected from synonymy: *inflatus* (Provancher) and *patulus* (Viereck). Two Holarctic species are redescribed: *melanurus* (Roman) and *minutulus* (Thomson). Ten Nearctic species are redescribed: *auriculiferus* (Viereck), *bicolor* (Lundbeck), *maculatus* (Provancher), *monticola* (Dalla Torre), *mucronatus* (Provancher), *paludicola* (Brues), *pubescens* (Provancher), *rotundiceps* (Provancher), *subclavatus* (Say), and *texanus* (Cresson). Sixty-three (63) species are newly described: *albior, albitexanus, angularis, arizonae, arkansensis, aurantifex, aurarius, aureolus, aurigena, auriger, bicolorescens, brachyceratus, brevicornis, callidius, callistus, chiricahuanus, chrysoleptus, concavus, coriaceus, daschi, declivis, durangensis, elegantulus, euryops, flavissimus, flavivittatus, gracilis, granulifacies, hesperus, hexamerus, julianus, latissimus, leioleptus, leopardus, leptotexanus, leucocnemis, melanogaster, michiganensis, nemati, nigrans, obscurus, occipitis, oregonianus, pentacrocus, pinidiprionis, praerotundiceps, pseudocallistus, punctatior, rhyssotexanus, rubescens, rugiceps, rugitexanus, rugosus, santacruzensis, serratus, spicus, spinissimus, taiganus, tetratylus, tricoloratus, tyloidiphorus, xanthopyrrhus,* and *xanthostomus*. A lectotype and paralectotype are designated for *paludicola* (Brues). Two holotypes are determined: *Phygadeuon montanus* Cresson and *Phygadeuon texanus* Cresson.

A polarity of characters is discussed in the section on character analysis; *Cubocephalus* (Gelinae, Echthrini) is used as the outgroup. Phylogenetic relationships of Endaseina *s.s.* (Gauld and Holloway, 1983: 193)—*Endasys, Glyphicnemis, Amphibulus,* and *Coptomystax*—are discussed under phylogeny. These relationships are based primarily on features of the clypeus, appendages, wings, propodeum, and abdomen.

Distribution and biology are summarized from literature and specimen data. *Endasys* is a K-strategist parasitizing prepupae in subterranean cocoons of Tenthredinidae, Diprionidae, and Argidae in coniferous or mixed-hardwood forests. Taxonomic and nomenclatural histories of the genus are reviewed under Taxonomy. Keys to species are presented separately for males and females. Diagnostic color patterns of the males of each species, and generalized figures of important morphological characters of both sexes are illustrated. Collection localities of each species are shown on maps and summarized in the text. References list historical citations of species' names as well as literature cited.

INTRODUCTION

This is the first revision of Nearctic *Endasys*, a common ichneumonid parasite of sawfly prepupae in subterranean cocoons. *Endasys* is a Holarctic genus, collected wherever pine, spruce, fir, hemlock, birch, willow, linden, and oak grow. *Endasys* is one of the larger genera of the Gelini (= Phygadeuontinae), and one of the more important parasites of sawfly cocoons in hidden situations in the forest floor. In this revision 77 species are recognized, including 12 old, 2 resurrected, and 63 new ones. It is based on over 12,000 pinned specimens in North American collections, as well as some Palearctic specimens.

The purpose of this revision is to diagnose and describe new and old species and to provide keys to them along with illustrations and maps. Biological and host information is summarized from specimens and the literature, and a character analysis and phylogeny are discussed. A complete reference list includes those merely listing Nearctic species names as well as important taxonomic and biological works. This study expands the known range of *Endasys* into the tundra of northern Canada and montane Mexico.

Historically, the genus, unlike the species, is well defined morphologically, and there are few problems defining its limits or justifying the species it includes. There is some overlap with *Amphibulus* by some species of *Endasys*, but these similarities shed light on the phylogenetic history of the genera rather than obscure their distinctness. In order to more clearly diagnose and describe the species, a format more like that of Boucek's has been used. This format groups characters that are the same or of similar aspect, rather than describing each body part separately. Within each category (e.g., color or shape), characters proceed from head to abdomen. Specific search images are more easily and accurately compared between species rather than comparing all details of each body part. The diagnosis lists critical characters in order of relative importance. Variation is detailed in the description, but only alluded to in the diagnosis.

Data from material examined, except for primary type information, has been severely abbreviated to shorten the text. Host data from specimens could not be confirmed, but was consistent with published names; Appendix I lists hosts recorded from the literature and from labels Appendix II lists detailed paratype locality data. Maps show all collection localities. Listed are American and Mexican states and Canadian provinces, museums; and total numbers for specimens studied.

Etymologies are from Latin roots, unless specified as from Greek or other sources; Greek roots are written with Latinized spellings. Word combinations were made according to classical Latin or Greek rules.

METHODS

Specimens were studied using a binocular American Optical dissecting microscope under the light of a 75 watt incandescent bulb. A micrometer was inserted into the objective eyepiece for measuring. High-intensity bulbs are unsatisfactory because of glare, making both color and surface features harder to evaluate. Fluorescent lighting reveals surface features such as punctation more clearly than incandescent lighting, but interferes with color interpretation, which is very important in *Endasys*. Although both 30- and 45-power were used in examining specimens, 30 allows greater evaluation of a specimen as a whole, while providing enough magnification to evaluate surface features such as punctation or flagellar tyloids.

Drawings were done either freehand, by tracing a template, or with a camera-lucida attachment. I used a standard drawing pen (rapidograph) with India ink on #201 Bristol board. The outlines of the scape, clypeus, front and middle legs, and the abdomen (= gaster, metasoma) were made by tracing a generalized template onto the Bristol board. The clypeus and abdomen were then modified to characterize the shape of each species. The holotype (male) of each new species was used to illustrate the color pattern, and the shape of the 1st and 2nd abdominal terga, and to draw the hind leg, using the camera-lucida attachment. For existing species, I used either a male paratype, a male compared with the type (homotype), or a male specimen chosen for illustration. Thus only males were selected to illustrate species.

Descriptions of new species are based upon holotypes, with variations reflecting paratypes. Only male specimens most like the type were labelled paratypes. Only females clearly associated with and similar to the (male) type are labelled paratypes. All other material identified as the species, but not used for the description, is listed as "Other Material Studied." Redescriptions of existing species are based on available primary types and specimens like it, often including secondary types, and specimens compared with types (homotypes) by Townes or me. Original descriptions were of little use in redescribing existing species.

Measurements and ratios used in diagnoses and descriptions are based on at least 5 specimens, usually more for very long series. The holotype and female paratypes are included for new species. For existing species, descriptions include available primary and secondary types, as well as specimens compared with the type. Generally, measurements are difficult to make because most surfaces are not flat.

Body lengths of males and females are approximated because specimens are rarely in a natural position on pins. They were measured from the anteriormost point on the head to the apex of the abdomen. This usually entailed one measurement from the head to the abdomen, and a second, of the abdomen from base to apex. The ovipositor or 3rd valvulae are not included in female lengths.

Other measurements were made as follows: *Hind femur* ratio is the maximum width divided by the maximum length of the femur including the trochantellus. *Radial cell* (front wing) length is length across the longest distance. *2nd discoidal cell* length is the longest distance diagonally across the cell. *Postpetiole* length is measured from the spiracle (center) to the lateral apex (not medially); the width, across the lateral apices. *Frons* ratio is the minimal distance between the compound eyes divided by the height of the frons from base of the scape to the clypeal apex; the height of the clypeus is included because it is difficult to determine the fronto-clypeal groove.

TERMINOLOGY

Most of the terms used in keys and descriptions follow Townes (1970: 36-48), with the following exceptions: trochantellui = 2nd-trochanters; area dentipara not 2nd lateral area; tergum(a) not tergite(s); sternum(a) not sternite(s); 3rd valvula not ovipositor sheath in the keys, diagnoses, and descriptions. In addition, the following descriptive terms are defined here for use in the keys, diagnoses, descriptions, and illustrations.

Clypeal margin (Fig. 78:A-E) ranges from "not upturned" to "strongly upturned"; "more or less" means sometimes more and sometimes less, usually applied to weakly upturned margin.

Male flagellum (Fig. 80:A-D) is "moderately slender" if the segments on its apical half are longer than wide; "moderately stout" if slightly longer than wide or square; and "stout" if square to wide.

Basal 3 flagellomeres of the female flagellum (Fig. 80:G-I) range from moniliform to distinctly elongate; the intermediate state is "short."

Glumes (Fig. 88:A-C) are "dense" if less than the narrow width apart, "moderately dense" if more than 1 but less than 2 widths apart, and "sparse" if more than 2 widths apart (Fig. 88:A-C).

Punctation is "dense" if pits are 1 diameter or less apart in all directions, "sparse" if more than 1 diameter in all directions, "impunctate" if without setiferous punctures; "moderately" means "less," "smooth and shiny" applies to the surface regardless of setiferous punctures, "more or less" means a mixture of punctures, and "variably" and "variable" mean the punctation is irregular.

Setiferous punctures (pits) (Fig. 86:A-F) are described according to relative width and depth: "coarse" means deep and closely spaced; "distinct" means deep with sharp edges; "more or less distinct" means variably deep or shallow with variably sharp edges; "indistinct" means shallow; "fine" or "finely distinct" means pit is slightly wider than its seta, with sharp edges; "smooth" means setiferous pit is as wide as seta; "rugulose" describes wrinkled surfaces with or without setiferous pits, although dense, coarse punctation is always rugulose.

Area dentipara (Fig. 83:A-C) is "regular" when shape is evenly trapezoidal; otherwise it is "narrowed" (shortened) or "wide" (lengthened).

Radial cell of the front wing is described according to its maximum length.

5

Radial sector (Rs) shape (Fig. 82:A-D) ranges from long, "bowed," or "straight,", to "short and curved".

Hind femoral ratios (Fig. 79;A-E) are evaluated differently for males and females. Male "swollen" = female "moderately swollen", and female "moderately slender" = male "moderately swollen" or "moderately slender".

CHARACTER ANALYSIS

Among the Gelini, *Endasys* exhibits a number of specialized characters due in part to the ground-hunting habit of the female. Price (1972a: 190) noted that female ichneumonids, including *Endasys subclavatus* (Say), that search for host cocoons on the forest floor had stouter flagella, stouter legs, and shorter abdomens and wings compared to ichneumonid species attacking larvae on plants. Within the genus there are several evolutionary trends away from the primitive endaseine ancestor. I have chosen the echthrine (= hemigastrine) genus *Cubocephalus* as the primitive outgroup for comparison. I consider the tribe Echthrini to be the most primitive of the 3 in Gelinae. It is mostly Holarctic, hosts are mostly cocoons of tenthredinid sawflies. Morphologically, many genera resemble Ephialtinae with long ovipositors, wing venation, and somewhat elongate abdomens. *Cubocephalus* generally attacks tenthredinid cocoons hidden in plant material, plant stems, under bark, or in dead wood (Townes, 1970: 129).

Trends in character variation of *Endasys* are discussed below, with mention of species groups exemplifying particular features.

Flagellum. The shape of the flagellum (Fig. 80:A-I) has phylogenetic significance. More generalized groups have a longer, more slender flagellum. This is demonstrated within species in that the female flagellum is always stouter than the male's. Males are seen to possess primitive features which are often derived in the female. In males, the flagellum is perceived as more slender if all segments are longer than wide. Specialization of this condition includes either lengthening or shortening of especially those segments on the apical half; Fig. 80:A-D illustrates these conditions. With females (Fig. 80:E-I), only the basal 3 segments and the apical half of the flagellum reveal the relative state of the flagellum. The generalized condition is best seen in the Texanus Group. The flagellum is relatively long, narrowed apically, and the basal 3 segments are distinctly elongate. The specialized condition, segments shorter and flagellum linear to apex, is best seen in females of the Santacruzensis and Bicolor groups.

Tyloids. There is a tendency toward reduction in the number of tyloids. These sensory structures of males (Fig. 87:A-D) are found on flagellomeres 8 to 13 or 14. Most *Endasys* males have 2 tyloids always on flagellomeres 10 and 11; the additional tyloids are either proximal or distal of these two. However, most Texanus Group males have 3 more or less distinct tyloids on segments 10-12. Other groups also have species with 3 or more of these placode sensillae. Females lack tyloids.

Annulus. A feature sometimes present on the female flagellum is an annulus on flagellomeres 4 or 5 to 9. The primitive state is a distinctly white, as seen in most females of the Texanus Group. Some species of the Bicolor, Auriculiferus, and Santacruzensis groups also have a distinct annulus, but it is often reduced to fewer segments. More commonly the flagellum is bicolored, with the basal half orangish or yellowish and the apical half black, at least dorsally. All species in the Subclavatus Group except *hesperus, albior,* and *nemati* have bicolored flagella, and several species in other groups have this character. The other common color state of the flagellum is unicolored, generally seen in many western and southwestern species as well as in a few of the Texanus Group. Some species exhibit a tricolored aspect, with the basal 4 or 5 segments orangish, segments 4 or 5 to 9 pale yellowish or whitish, and the apical half black.

Clypeus. The clypeus (Fig. 78:A-E) becomes more impressed apically as the clypeal margin becomes sharper and distinctly upturned. The primitive, convex state with the margin not or weakly upturned is seen in the Texanus Group, in many members of the Subclavatus Group, and in a few of the Monticola Group. The generalized shape of the clypeus is broadly oval, less than 3 times wider than high. The derived condition is seen in the Bicolor Group, where the clypeus is 3 or more times wider than high, with a more or less upturned margin.

Vertex. The lengthening and flattening of the vertex (Fig. 89:B) is probably a derived condition for the genus. There are tendencies toward this condition in all species groups. It is best observed in *occipitis* and *inflatus.*

Hind Leg. The color of the hind femur and tibia have considerable diagnostic value. I have determined that the primitive color is black, sometimes with whitish on the tibia dorsally, as seen in some species of the Texanus Group. The derived condition is considered to be a reduction of black to the apices, or its absence. Species with usually orange femora may sometimes occur with black apically, though species that are typically with black are never completely orange. Species with all-black femora may occasionally have it apically and suffusions dorsally or laterally. The Monticola Group best demonstrates these variations.

The shape of the hind femur is a useful diagnostic tool, and can be used to infer evolutionary trends. The relative slenderness or stoutness is usually accurately perceived by the eye without measurement. However, it can be quantified most of the time by a femoral ratio found by dividing the maximal depth by the length including the trochantellus; Fig. 79:A-E shows characteristic femora and their ratios. Most male *Endasys* have the generalized state, moderately slender or moderately stout. Females generally have stouter femora than males, presumably due to their habit of groundhunting for hosts. A few species show a distinctly slender hind femur, which may or may not be a secondary apomorph within the genus. Females of these species also have distinctly slender femora, which would seem to signal the plesiomorphic condition.

Propodeum. The propodeum is difficult to interpret. However, the relative slope and the quality of the carinae and the apophysis are diagnostically useful for some groups. The generalized condition for the propodeum, seen in the Texanus Group, appears to be dorsally level, coarsely carinate, and with a strong and toothlike apophysis. Regarding the carinae and apophysis, the generalized condition is actually specialized compared to the outgroup *Cubocephalus*, which has indistinct carinae and no apophysis, but does have a rather level propodeum. Species having a propodeum with reduced carinae and apophyses exhibit many other specialized features, so I have judged these to be reductions of a secondary nature. The levelness is best judged laterally, observing the general line of the dorso-lateral carina leading to the apophysis. It is more easily observed in females, in which the carinae are finer and the surfaces smoother than in the males. The Santacruzensis Group best shows the derived condition.

Two dorsal areas are helpful in evaluating the propodeum: the relative shape of the hexagonal areola and that of the trapezoidal area dentipara (area with the apophysis). The generalized condition of the areola is broadly to widely hexagonal (appearing rectangular) (Fig. 85:C,D), best seen in the Texanus and Subclavatus groups; that of the area dentipara is narrow (Fig. 83:C). In the derived condition, the areola and the area dentipara are lengthened (females Fig. 85:A, males Fig. 85:B). The Santacruzensis Group typifies this condition, as do some species of the Monticola and Auriculiferus groups. There is considerable interspecific variation, making these characters of limited diagnostic value except within small assemblages of species. Generally, as the propodeal carinae become reduced these areas become lengthened.

Wing. Useful phylogenetic features of the wings are: degree of darkening, shape of the radial cell, and sometimes the interception of the brace vein by the nervulus. Darkening of the wings appears to be a geographical trait, common in western North American species. Wings of eastern and southwestern species are generally hyaline. The generalized condition is slightly darkened or tinted, as seen in the Texanus Group.

The radial cell (Fig. 82:A-D) is primitively long, radial sector (Rs) straight, and slightly bowed (Fig. 82:A). The bowing would presumably be the continuation of Rs to the apex of the wing margin. In more primitive species, both males and females have radial cells of similar shape and length, as seen in the Texanus Group. Most species in the other groups show a shortening and then curving of the female Rs. This character is useful in associating the sexes.

The nervulus (a cu-a cross-vein) is typically distad of the brace vein (a m-cu cross-vein). This is considered plesiomorphic in *Endasys*. Several species in the relatively specialized Auriculiferus and Santacruzensis groups have the nervulus near the brace vein, distad by about twice the width of the vein.

Abdomen. The relative shape, stoutness, and distinctness of the dorsomedian carina of the 1st abdominal tergum (Fig. 84:A-G) in both males and females is an important character. The generalized state is moderately stout, with the postpetiole squarish in the males (Fig. E-F) and gradually widened in the females (Fig. 84:B). The carinae are strong and conspicuous beyond the spiracle. The specialized condition is best seen in the Santacruzensis and Auriculiferus groups. Typically, as the postpetiole becomes narrow, the petiole becomes longer and more slender and the sternum lengthens to or beyond the petiolar spiracle. In the generalized state the sternum of the petiole is always before the spiracle.

In males, the extent of sclerotization of the 5th and 6th abdominal sterna has phylogenetic value (Fig. 81:A-C). The 5th sternum is rather variable within species, although when it is complete medially (that is, sclerotized) it is rarely membranous medially, at least not apically. In more generalized groups, such as the Texanus and Subclavatus groups, it is typically complete, and the 6th is rarely membranous medially. In the specialized Santacruzensis Group both 5th and 6th are commonly membranous medially. This character is often difficult or impossible to evaluate because of wrinkling of the ventral aspect of pinned male *Endasys*. Females are membranous medially as far as the subgenital plate, showing again a generally derived condition compared with the males.

Ovipositor. The length of the 3rd valvula (ovipositor sheath) relative to the length of the hind tarsomeres is of some diagnostic and phylogenetic value. The generalized condition is moderately long—about as long as the basal 4 hind tarsomeres, but always longer than the basal 3, and not longer than all 5. One problem in evaluating this character is that larger specimens appear to have a shorter 3rd valvula than smaller specimens of the same species. It should be noted that the 3rd valvula is always as long as the part of the ovipositor that extends beyond the apex of the abdomen, so that its length should be correlated with that of the sheath. This extended length of the ovipositor can be used as a measure of the 3rd valvula when the latter is curled, broken, or missing. The length of the ovipositor is not used because most of the time it is not entirely visible.

Abdomen Color. Abdominal color is very useful in diagnosing species, but appears to have no definitive patterns with regard to species groups. That is, most species groups have several abdominal patterns represented. The generalized color would seem to be black, with all others derived. There appear to be two basic trends in the reduction of black. More commonly, black recedes to the apex of the abdomen, making it "bicolored," commonly seen in the Bicolor and Monticola groups, but also occasionally in the Texanus Group. The other trend is for black to recede anteriorly, as in some species of the Subclavatus and Santacruzensis groups. All intermediates are seen, particularly in males. Females are less distinctly colored, and the usual color for most species is entirely orange or bicolored. Few species of *Endasys* have entirely black abdomens.

Color. The presence of white coloration on the scape, clypeus, tegula, coxae, trochanters, and front and middle tibiae dorsally has diagnostic and phylogenetic value. Most females do not show white on these areas even when the male of the species does. However, species in the Texanus Group often have distinct white on the front and middle tibiae dorsally of both males and females. *E. maculatus* and *mucronatus* also have white on the hind tibia dorsally. White in males becomes either more extensive or reduced. Typically those species with a white or yellow scape and clypeus also have white on the other areas. The Subclavatus Group is characterized primarily by the presence of white especially on the coxae and trochanters of males. In Texanus Group males, white appears typically on the scape apically, tegula, coxae apically, and trochanters ventrally. Most males have white on the front and middle tibiae dorsally, even if white is reduced or absent on other areas. White on males of the Santacruzensis Group is as extensive as in the Subclavatus Group, except the tibiae are more uniformly yellow or orange, usually without pale areas dorsally. In males of the Monticola and Bicolor groups, black, orange, or yellowish replace white.

It is important to distinguish between yellow areas that are not pigmented with white from areas where white is actually present, but appears pale yellowish because of the thickness of the integument. Typically, pigmented-white areas are not translucent, and muscles cannot be seen through the integument. When the integument is yellow (or orange) due to the lack of melanization, it appears glassy and translucent. Generally, white on the legs is conspicuous and varies from ivory to pale, yellowish white. White on the scape and clypeus varies from ivory to yellow, but the yellow never merges into orange. Orange, however, can range from yellowish orange to entirely black.

Punctation. On the head and on the thoracic pleura the general pattern of punctation has diagnostic value. The generalized condition seems to be evenly dense punctation with distinct, often conspicuous pits, with the surface more or less rugulose. The Texanus Group again most clearly demonstrates this aspect. Some females of the Subclavatus and Auriculiferus groups approach the primitive condition. Exemplifying specialized punctation is the Santacruzensis Group, with fine, more or less indistinct punctation, and surfaces usually smooth and shiny.

ENDASYS SPECIES GROUPS

The genus can be divided into 6 groups based primarily on clypeal shape; form of the clypeal margin; spacing and distinctness of pits, particularly on the head; coarseness of propodeal carinae; distinctness of carinae on 1st abdominal tergum; presence of white on appendages; and the shapes of the 1st tergum, hind femur, and female flagellum. Several other characters are used together with the above to group species. There are practical difficulties, however; in precisely defining each of the species groups; thus no key to them is presented.

Characteristic features of each species group are presented before the species descriptions, which are given in alphabetical order. Other remarks on species groups are made in the Character Analysis and Taxonomy sections. All 6 species groups are represented in the Palearctic (Luhman and Sawoniewicz, 1991). The following lists the species included in each species group. The groups are listed in phylogenetic order (plesiomorphic to apomorphic), while the species in each group are listed alphabetically.

Texanus Group: *albitexanus, angularis, latissimus, leptotexanus, maculatus, mucronatus, oregonianus, pseudocallistus, rhyssotexanus, rubescens, rugitexanus, rugosus, taiganus, texanus, xanthopyrrhus, xanthostomus.*

Subclavatus Group: *albior, aurarius, aurigena, brevicornis, chrysoleptus, hesperus, inflatus, michiganensis, nemati, paludicola, patulus, praerotundiceps, pubescens, rotundiceps, subclavatus, tyloidiphorus.*

Monticola Group: *bicolorescens, brachyceratus, daschi, hexamerus, leioleptus, leopardus, melanurus, monticola, obscurus, pentacrocus, serratus.*

Bicolor Group: *bicolor, callistus, coriaceus, declivis, euryops, minutulus, nigrans.*

Auriculiferus Group: *arkansensis, aurantifex, auriculiferus, elegantulus, granulifacies, rugiceps, spicus.*

Santacruzensis Group: *arizonae, aureolus, auriger, callidius, chiricahuanus, concavus, durangensis, flavissimus, flavivittatus, gracilis, julianus, leucocnemis, melanogaster, occipitis, pinidiprionis, punctatior, santacruzensis, spinissimus, tetratylus, tricoloratus.*

PHYLOGENY

Endasys Foerster is closely related to the Holarctic genera *Glyphicnemis* Foerster and *Amphibulus* Kriechbaumer and the Oriental *Coptomystax* Townes. Within the subtribe Endaseina all 4 share the apomorphies of the transverse break on the posterior margin of the mesoscutum (Gauld and Holloway, 1983) and a ridge across the prescutellar groove (ridge often reduced or absent in *Amphibulus*). The remaining genera of the subtribe, lacking the transverse break and the ridge, are more distantly related.

Endasys, Amphibulus, and *Glyphicnemis* each share several apomorphies (Luhman, 1986). *Endasys* and *Glyphicnemis* share a strong median ridge across the prescutellar groove, autapomorphic in Gelini and most Ichneumonidae except Xoridinae; pronounced dimorphism, apomorphic in Ichneumonidae; and stouter body and appendages, apomorphic in Endaseina. However, *Endasys* and *Amphibulus* lack the elongate lower mandibular tooth and the subapically inserted tibial spurs, both autapomorphic in *Glyphicnemis*. On the other hand, *Amphibulus* and *Glyphicnemis* both have a more modified 1st abdominal tergum (more slender and elongate; dorsomedian carina weak or absent in most species) and propodeum (more elongate; apophysis either lacking, or spike-like in many *Amphibulus*).

Overall, *Endasys* retains more plesiomorphies than the other two. This is especially shown by its generally less modified wing venation (radial cell elongate, Rs straight or bowed), clypeus (usually weakly or not upturned, never with teeth or lobes), appendages (usually moderate proportions), propodeum (apophyses rarely absent), and 1st abdominal tergum (usually moderate proportions; dorsomedian carina distinct). This suggests that *Endasys* reflects the more plesiomorphic state of Endaseina (*s.s.*). Within the genus, the Texanus Group best demonstrates plesiomorphic states; the Santacruzensis and Auriculiferus groups, the apomorphic states.

Within the species groups of *Endasys*, the Palearctic and the Nearctic appear the same morphologically, although the Palearctic species show a more generalized condition overall. For example, the male's flagellum often has 3 tyloids, the female's often a white annulus; the anterior abdominal terga of males are often mat or distinctly punctate; and coloration is generally blacker, the abdomen commonly more or less bicolored—blackish on the apical terga.

Several sister-species relationships are observable with western Palearctic species in every group: *maculatus* (Provancher) with *parviventris* (Gravenhorst), *texanus* (Cresson) with *cnemargus* (Grav.), *inflatus* (Prov.) with *testaceus* (Taschenberg), *subclavatus* (Say) with *annulatus* (Habermehl), and *obscurus* and *gracilis* each with a new species in Europe. Additionally, there are strong similarities and possible relationships between the following: *serratus* with *brevis* (Grav.), *xanthopyrrhus* with *amoenus* (Habermehl), *pubescens* (Prov.) with *euxestus* (Speiser), *latissimus* with *analis* (Thomson), *auriculiferus* (Viereck) with *nitidus* (Haberm.), and several Santacruzensis Group species with *talitzkii* (Telenga). Three Holarctic species have been determined: *minutulus* (Thomson), *melanurus* (Roman), and *coriaceus*. As the eastern Palearctic *Endasys* are studied, perhaps more Holarctic species and more Palearctic sister-species will be determined.

From the widespread distribution of the species groups, it would appear that they had evolved before the Beringia land bridge allowed dispersal between North America (Alaska) and Asia (Siberia) during the Tertiary (Matthews, 1980:1092). Inferences on *Endasys* dispersal can be based on fossil pollen evidence, as summarized by Matthews for Canada (1979; 1980). It shows that coniferous forests reached most of northern Canada and Greenland from mid-Miocene until about mid-Pliocene, when much of this region became tundra. The effect of Pleistocene glaciation and climatic change acted mainly to reduce and disrupt the distributions of the forests, as well as the insects in them, already existing since the Miocene (Matthews, 1979: 58; 1980: 1089-1090). This view seems consistent with the number of Holarctic sister-species—that is, species between realms are more closely related to each other than those within realms, suggesting that the most recent speciation has been due to vicariance as a result of the final submergence of Beringia in the Pleistocene. So it would appear that Endasys has persisted as a relatively conservative group associated with a forest environment, having reductions and expansions since at least the Miocene.

BIOLOGY

Endasys is Holarctic, and in the Nearctic has been collected principally where Pinaceae occur. The distribution of *Pinus, Picea*, and *Tsuga* corresponds generally to that of *Endasys*. Thus *Endasys* is collected in a wide range of latitudes and altitudes. Among material examined were specimens from the mountains of central and northern Mexico, Santa Cruz Island, California; Mt. Rainier, Washington; Mt. McKinley, Alaska; and from the 10,000-ft. level in Colorado. Some species, although occurring in coniferous forests, parasitize argid sawflies on *Betula* (birch). A few other species occur in tundra regions, and many eastern species occur in mixed deciduous-coniferous forests, or in areas with few conifers. In these cases, the host rather than the host-plant is more important. Therefore *Endasysis* more predictably described as a parasite either in pinaceous forests, or in forests or tundra that are host to sawflies of the Argidae or nematine and heterarthrine Tenthredinidae. The argid habitats would be primarily *Betula* (birch); the Nematinae, *Salix* (willow), *Populus* (poplar) or tundra vegetation; and the Heterarthrinae, *Quercus* (oak) or *Betula*. *Endasys* has not been collected in tropical, desert, or grassland areas.

Based on reared specimens and published records, *Endasys* is an external parasite of sawfly prepupae in cocoons, and so is called a cocoon parasite. Studies by Furniss and Dowden (1941), Price (1972b), Rau (1976), and Thompson and Kulman (1980) have shown that the females only parasitize sawfly cocoons hidden in the duff or under other forest litter, and not cocoons on foliage or under bark. Host tenthredinoids for *Endasys* include the Argidae, Diprionidae, and Tenthredinidae (Nematinae and Heterarthrinae). Price (1971, 1972a) determined that *Endasys* is a common ichneumonid in the sawfly cocoon parasitoid guild. He dubbed this group K-strategists (1974a, 1975) because they laid fewer eggs in the host stage having the lowest mortality. Larval parasites, the r-strategists, laid many more eggs to compensate for considerably higher host mortality and, therefore, parasite mortality.

Within the cocoon-parasite guild, *Endasys* specializes in cocoons buried in the duff on the forest floor. The females have strong spines on the anterior faces of the front and middle tibiae which are effective in digging through such material. Thus they effectively compete with other geline species which can only attack cocoons close to the surface. Experiments by Thompson and Kulman (1980), using drop-traps to collect 5th or 6th instar larvae of *Pikonema alaskensis* (Rohwer), showed that *E. pubescens* (Provancher) can dig through a sand-sawdust mixture to parasitize the sawfly cocoons. This ground-hunting habit of the female is reflected in the morphology of the genus generally. Price (1970a, 1972a) observed that ground-hunting female ichneumonid cocoon-parasites, including *E. subclavatus* (Say), have shorter wings, stouter legs and antennae, impectinate tarsal claws, and longer ovipositors, compared to those attacking larvae in exposed situations on stems or foliage.

The specialized habit of the female has resulted in sexual dimorphism generally throughout the genus, including color patterns. Differences are discussed in the Character Analysis section, and have necessitated keys for each sex. Males are typically more slender and more conspicuously colored than females. White coloration of the scape, clypeus, coxae, and trochanters are not common on females. A conspicuous feature of several species (only females) is a white annulus medially on the flagellum. Females, and some males, of other Gelinae and Ichneumoninae (also ground-hunting parasites) typically have these annuli. This feature is noticeably rare or absent in all other subfamilies of Ichneumonidae. It is possible that the annulus is connected with mimicry of hunting wasps such as sphecoids and pompilids. I have seen such annuli on the antennae of Coleoptera, Diptera, and Hemiptera mimicking wasps. The body coloration of *Endasys* females, basically black head and thorax with an orange or bicolored abdomen, reflects that commonly seen in sphecoids. Females lacking a distinct annulus have the flagellum bicolored or tricolored as viewed dorsally, or it is unicolored orangish or blackish.

The only apparent effect of parasitizing subterranean sawfly cocoons on *Endasys* male morphology is the presence of spines on all the tibiae dorsally, and well-developed tibial spurs, longest on the hind tibia. These probably aid the male in digging through loose duff after emerging from the cocoon. The hairy integument of both sexes may protect them from abrasion during emergence from the soil.

The phenology of the genus shows some correlation with host phenology. Bobb (1963: 620) reports that *E. subclavatus* takes about 6 weeks to develop from egg to cocoon. It overwinters as a larva in its cocoon within the host cocoon. Parasite emergence begins in the spring and continues until fall. Adult emergence peaks in late June and again in September. Adult parasites are present throughout the time that prepupal hosts are in cocoons. Collected material for most species spans May through September, with most specimens June through August. In southern California, the southwestern United States, and montane Mexico, collections are generally from March to May. Males emerge about 2 weeks before females. Studies by Rau (1976) showed 1 generation a summer, the emergence of *E. pubescens* peaking 2 to 3 weeks after emergence of adult *Pikonema alaskensis*, but at or before host larval drop. A study of *Euceros* (Ichneumonidae) by Tripp (1961: 53) revealed that *Endasys subclavatus* had "at least" 2 generations per year, with a spring and an autumn emergence. Thus the parasite attacks the overwintering host generation in the spring and the host's progeny in the autumn. A study by Price (1972b) also showed 2 generations of *E. subclavatus*, with the 1st-generation emerging at or before emergence of adult *Neodiprion swainei* Middleton, and the 2nd generation in autumn when the host is in its cocoon. He also observed that only the 2nd generation produced many progeny. He proposed that this was possibly the result of increased hyperparasitism by other ichneumonid parasites on the remaining overwintering sawfly cocoons parasitized by 1st generation (spring) *subclavatus*.

Mullier (1979: 25) observed that female *subclavatus* lives about 33 days (+/-4) and lays about 22 eggs (+/-8). This longevity points to the possibility of parasitism of both host generations in one season. These observations emphasize the specialized niche of *Endasys*—namely that it attacks suitable host cocoons in the duff or soil, regardless of age. It would be interesting to know if pupal cocoons are also attacked. Unsuitable hosts could explain the high mortality of 1st-generation *subclavatus*.

Host specificity in *Endasys* has not been established, but rearings (CNC, UM, FLDA, OSU) from plantations and forests in the United States and Canada yield predictable complexes of parasites of all guilds from a sawfly species. For example, rearings of *Pikonema alaskensis* from plantations of *Picea glauca* (white spruce) in the Grand Rapids area of Minnesota from 1973 to 1980 (Houseweart and Kulman, 1976; Rau, 1976; Thompson and Kulman, 1980) consistently yielded the parasites *E. pubescens*, a dozen other species of ichneumons, a braconid, a pteromalid, and a tachinid. Nevertheless, this species has been recorded from 2 families and several genera of tenthredinoids. There are multiple host records also for *patulus* (Viereck) and *subclavatus*, although all of these latter are misidentifications of the former. Thus it is likely that other species of *Endasys* parasitize more than one species of sawfly, depending on host size, seasonality, and forest habitat.

There is little other biological information on *Endasys*. Iwata (1960: 158) illustrates the ovarian egg, egg, and ovary of a Palearctic species. Short (1959) and Finlayson (1960) illustrate the cephalic structures and spiracles of the last larva instar.

TAXONOMY

Synonymies of the genus are based on Townes (1970: 83), Carlson (1979: 417), and a study of the type species of the Foerster genera by Perkins (1962: 451).

Genus *Endasys* Foerster, 1868

Endasys Foerster, 1868: 184. **Type species**: *Stylocryptus analis* Thomson; included by Roman, 1909: 243; designated by Viereck, 1914: 139.

Scinascopus Foerster, 1868: 185. Type species: *Phygadeuon parviventris* var. *cnemargus* Gravenhorst, by subsequent monotypy from inclusion by Brischke, 1891.

Bachia Foerster, 1868: 186. Preoccupied by Gray, 1845. Type species: *Phygadeuon (Bachia) testaceipes* Brischke, by subsequent monotypy from inclusion by Brischke, 1891.

Stylocryptus Thomson, 1883: 520-521. Type species: *Phygadeuon brevis* Gravenhorst, by monotypy.

Bachiana Strand, 1928: 52, new name for *Bachia* Foerster.

The Palearctic *Endasys* were first placed in *Phygadeuon* Gravenhorst (1829: 635), his Familia III, within "Cryptus V.," a Gravenhorst "Genus" that is the equivalent of a subfamily now. The "Familia" *Phygadeuon* was characterized by the squarish, broad head, shortened antenna, areolet closed and 5-sided, abdomen subsessile and a little petiolate, and ovipositor exserted.

Foerster (op. cit.) described 3 different genera for what is now *Endasys*. These coincide with 3 of 6 species groups. *Endasys* represents the Monticola Group; *Scinascopus*, the Texanus Group; and *Bachia*, the Subclavatus Group.

Thomson (op. cit.) erected *Stylocryptus* for the 3 Foerster genera, apparently recognizing the characteristic synapomorphy of the bipartite prescutellar groove. *Glyphicnemis* (=*Gnathocryptus* Thomson), the other genus sharing this character, was separated from *Endasys* in his key by the denticulate tibiae and the elongate lower mandibular tooth.

Habermehl (1912: 165; 1916: 376) placed both, reduced to subgenera, within Thomson's *Stylocryptus*. Since the subgeneric names were never used, all subsequent

18

Palearctic species of these genera were referred to as *Stylocryptus* until Townes (1944: 213-217) renewed the generic rank of *Endasys*, divided into 2 subgenera: *Endasys* and *Glyphicnemis*.

Most Nearctic species of *Endasys* were placed in *Phygadeuon* until Cushman (1922: 28; 1925: 389; 1928: 928) placed several names in *Stylocryptus*. Nevertheless, the use of *Phygadeuon* continued in Nearctic literature (primarily host and locality lists) until Townes' 1944 catalogue. Other generic names under which species of *Endasys* have been described are *Cryptus, Ichneumon, Alomya, Platylabus, Bathymetis, Medophron,* and *Oxytorus*. Townes in his generic treatment of the world Ichneumonidae (1970: 83-84) described *Endasys* and *Glyphicnemis* as separate genera. After 1940, *Endasys* predominated in the literature as well as on labels.

The diagnostic characters of the genus today are the same as those recognized over a hundred years ago. The following (modified from Townes, 1970: 83-84) characterizes the Nearctic *Endasys*.

Front wing 2.7-8 mm long. Body 2.5-10 mm long. Female body stout; male, stout to slender. Head big, transverse, subrectangular in dorsal view. Upper margin of face concave, with or without a very small, median, rounded tubercle. Genal carina joining oral carina at mandible. Clypeus (Fig. 78:A-E) rather small and elliptical, sometimes widened, in profile almost flat, its apical margin convex or concave, thick and squared off, margin upturned or not. Flagellum (Fig. 80:A-F) stout to moderately slender. Male usually with 2 or 3 tyloids on flagellomeres 9-12; female flagellum unicolored, or pale on segments 1-9, or with white (annulus) on segments 4-9. Lower tooth of mandible slightly to much shorter than upper tooth. Eye surface with hairs, these usually short and sparse, but sometimes long and conspicuous. Prepectal carina ending dorsad at subtegular ridge. Mesoscutum weakly to moderately convex. Notaulus present or absent, not reaching to center of mesoscutum. Posterior edge of mesoscutum with a transverse break that is rather narrow and weak, laterally evanescent. Prescutellar groove usually with a strong median longitudinal carina. Areolet pentagonal and closed, slightly narrowed to slightly widened. Nervulus distad of brace vein. Nervellus inclivious, intercepted. Tibial bristles weak to moderately stout, stouter in females. Apical truncation of hind tibia approximately transverse. Spurs of hind tibia inserted approximately at apex; longer spur about half as long as hind basitarsus. Areola (Fig. 85:A-D) widely to elongate hexagonal, apophysis usually distinct, but may be reduced or absent; area dentipara (Fig. 83:A-C) trapezoidal or nearly triangular. First tergum (Fig. 84:A-G) moderately stout to moderately slender, its dorsomedian carina strong, at least in front of spiracle, stronger in males. Second tergum of males narrowed or widened; females' always widened. Terga 2 and 3 usually polished in female, polished to mat in male; moderately hairy in male, almost hairless in female. Epipleura of terga 2 and 3 narrow, folded under. Sterna 4 and 5 of male complete medially or not, sternum 6 usually incomplete; sterna of female membranous to subgenital plate. Male subgenital plate strongly to weakly emarginate apico-medially. Ovipositor sheath (3rd valvula) shorter than basal 2 hind tarsomeres to 1.5 times as long as hind tarsus, usually shorter than hind tarsus. Ovipositor of moderate stoutness, compressed, its tip elongate lanceolate, ridges present but not distinct.

The genus is always separable from *Glyphicnemis* by the more apical insertion of the tibial spurs, not denticulate as with the latter, and the upper tooth of the mandible as long as or longer than the lower tooth. Most *Amphibulus* are separable by the absence of the ridge across the prescutellar groove, and a distinct transverse break across the mesonotum in front of this groove. Several Nearctic species of *Amphibulus* approach *Endasys* by the presence of a ridge in the prescutellar groove and a less distinct transverse break across the mesoscutum. These can be separated from *Endasys* by the genal carina joining the oral carina distinctly before the mandible, the sternaulus coarser anteriorly, and posteriorly distinct across the mesopleurum, and the inner hind tibial spur shorter—about 0.3 times as long as the hind basitarsus or less. In addition, *Amphibulus* has the clypeal margin sharper and always upturned, sometimes toothed or bilobed; the 3rd lateral area is often defined; apophysis usually toothlike or absent; appendages and petioles of both sexes are more slender; nervulus usually intercepting or slightly distad basal vein; flagellum of males usually with 3 or more tyloids; and flagellum of females (except three species) and some tropical males with white annulus.

KEYS

Male and female keys are presented for the Nearctic species. The keys are intended to be artificial, although they occasionally reflect phylogenetic groupings. Characters (Figs. 78-89) used in the keys and descriptions are discussed in the section on Character Analysis; measurements and terminology are explained in the Methods and Terminology sections.

Nearctic *Endasys* Foerster

Males

1a.	Coxae and trochanters partly or entirely white (Figs. 1:C, 6:C, 8:C, 18:C, 24:C, 37:C)... 2
1b.	Coxae and trochanters orange, yellow, or black (Figs. 3:C, 7:C, 9:C, 11:C, 40:C, 67:C)... 53
2a.	Thorax and propodeum orange...*aurarius* sp. n.
2b.	Thorax and propodeum black.. 3
3a.	3 tyloids... 4
3b.	2 tyloids... 16
4a.	Tyloids on flagellomeres 9-11...*auriculiferus* (Viereck)
4b.	Tyloids on flagellomeres 10-12... 5
5a.	Abdominal terga 2-6 orange.. 6
5b.	Abdominal terga 2-6 partly or entirely black... 12
6a.	Clypeus black... 7
6b.	Clypeus yellow or white... 9
7a.	Hind femur black (orange form)...*mucronatus* (Provancher)
7b.	Hind femur mostly orange.. 8

8a.   2 tyloids on segments 10-13, mesopleurum smooth and shining; Arizona............
        ....*tetratylus* sp. n.
8b.   3 tyloids on segments 10-12, mesopleurum rugulose; E. North America............
        ....*texanus* (Cresson)

9a.   Front and middle coxae mostly white, mesopleurum mostly smooth centrally........
        ....*tyloidiphorus* sp. n.
9b.   Front and middle coxae mostly yellow, orange, or black, white apically,
        mesopleurum rugulose centrally 10

10a.   Hind femur yellow-orange, slender, nearly parallel sided (Fig. 79:A), its ratio 0.18
        ....*leptotexanus* sp. n.
10b.   Hind femur orange with black apex, moderately slender (Fig. 79:B), ratio 0.21-
        0.22...11

11a.   Hind tarsomeres white, coxae without black.........................*albitexanus* sp. n.
11b.   Hind tarsomeres black, coxae partly black.........................*rugitexanus* sp. n.

12a.   (5.) Abdomen entirely black...13
12b.   Abdomen orange and black...14

13a.   Front and middle tibiae yellow; Arizona ....*melanogaster* sp. n.
13b.   Front and middle tibiae mostly black, tibiae white dorsally; E. North America.......
        ....*mucronatus* (Provancher)

14a.   Mesopleurum rugulose centrally, 2nd abdominal tergum orange, postpetiole with
        strong dorsomedian carina...*rugitexanus* sp. n.
14b.   Mesopleurum mostly smooth centrally, 2nd tergum partly black, postpetiole with
        weak dorsomedian carina...15

15a.   Clypeus white, postpetiole nearly square (Fig. 84:F); E. North America.............
        ....*tyloidiphorus* sp. n.
15b.   Clypeus black or orange, postpetiole elongate (Fig. 84:G); California................
        .... *santacruzensis* sp. n.

16a.   (3) Abdominal terga 2-6 orange or yellow...17
16b.   Abdominal terga 2-6 partly or entirely black...34

17a.   Abdomen, hind leg, scape, clypeus mostly yellow; Mexico........*flavissimus* sp. n.
17b.   Abdomen mostly orange, other features variably orange or black; U.S. or Mexico..
       18

18a. Hind femur mostly black... 19
18b. Hind femur mostly orange... 22

19a. Clypeus mostly black; Mexico .... *spinissimus* sp. n.
19b. Clypeus white or yellow; U.S.A or Mexico.. 20

20a. Clypeal margin not upturned (Fig. 78:A), area dentipara narrow (Fig. 83:C), propodeal carinae strong; Alaska to E. North America...... *pubescens* (Provancher)
20b. Clypeal margin distinctly upturned (Fig. 78:D,E), area dentipara regular or wide (Fig. 83:A,B), propodeal carinae weak but distinct; Mexico and southwestern U.S. 21

21a. Hind tibia yellow with black apex, frons rugulose with coarse, dense pits; Mexico .... *pinidiprionis* sp. n.
21b. Hind tibia black, frons with moderately dense to moderately sparse punctation, surface smoother; Arizona..*arizonae* sp. n.

22a. Clypeus black or orangish black, clypeal margin distinctly upturned (Fig. 78:D,E) 23
22b. Clypeus yellow, white, or orange, clypeal margin not or weakly upturned (Fig. 78:A,B) 24

23a. Head rugulose with very dense, coarse, punctation, hind coxa orange; E. North America .... *rugiceps* sp. n.
23b. Head with moderately dense punctation, surface mostly smooth, hind coxa blackish; Arizona.. *concavus* sp. n.

24a. Species occurring W. of the Rocky Mts. 25
24b. Species occurring E. of the Rocky Mts. 27

25a. Hind coxa orange, hind femur usually orange, sometimes black apically; Pacific Northwest...*hesperus* sp. n.
25b. Hind coxa partly black, hind femur always black apically; western U.S. 26

26a. Clypeal margin not or weakly upturned (Fig. 78:A,B), frons with evenly dense punctation, tegula yellowish or brownish, front and middle coxae partly blackish; Montana to Colorado .... *paludicola* (Brues)
26b. Clypeal margin distinctly upturned (Fig. 78:D), frons with variably dense punctation, tegula white, front and middle coxae without black; southern California. .... *tricoloratus* sp. n.

27a.   Hind tarsomeres whitish, 1st abdominal tergum orange and moderately slender (Fig. 84:F), hind femur usually entirely orange, hind coxa orange, moderately large to large, 7-9 mm long .... *subclavatus* (Say)

27b.   Hind tarsomeres blackish or orange, 1st abdominal tergum variable, often black with orange apically, hind femur usually black apically, hind coxa and size variable 28

28a.   Postpetiole elongate (Fig. 84:G), hind femur slender to moderately slender (Fig. 79:A,B), ratio 0.19-0.21, hind coxa partly black.................. *chrysoleptus* sp. n.

28b.   Postpetiole nearly square, hind femur more swollen (Fig. 79:C,D), ratio greater than 0.21, hind coxa with or without black... 29

29a.   Small, under 5 mm long, flagellum short and stout (Fig. 80:D), face finely rugulose, front and middle coxae white apically at most............ *brevicornis* sp. n.

29b.   Larger, over 5 mm long, flagellum moderately stout to moderately slender, face variable, front and middle coxae often mostly white 30

30a.   Middle coxa mostly orange or yellowish, hind coxa orange, hind femur and tibia orange with black apically, 1st abdominal tergum moderately stout (Fig. 84:E), dorsomedian carina distinct, 2nd tergum wide .... *aurigena* sp. n.

30b.   Middle coxa mostly white, color of hind coxa and femur variable, 1st tergum variable, often moderately slender (Fig. 84:F) 31

31a.   Hind femur slender (Fig. 79:A), ratio 0.20, flagellum long and slender (Fig. 80:A), tyloids long (Fig. 87:A), large, 7-85 mm long........................*patulus* (Viereck)

31b.   Hind femur moderately swollen or swollen (Fig. 79:C,D), flagellum stouter (Fig. 80:B,C), smaller, 5-7 mm long 33

32a.   (41.) First abdominal tergum nearly linear, hind femur slender, ratio 0.21 ....*flavivittatus* sp. n.

32b.   First abdominal tergum widened at postpetiole, hind femur moderately swollen, ratio   0.24... *santacruzensis* sp. n.

33a.   Clypeal margin sharp (Fig. 78:B), face finely granular with very dense, very fine pits, hind femur swollen (Fig. 79:D), ratio 0.25-0.28, propleurum mostly punctate with moderately dense pits, head with vertex slightly lengthened and flattened (Fig. 89:B) .... *inflatus* (Provancher)

33b.   Clypeal margin not or weakly upturned, face finely rugulose with very dense pits, hind femur moderately swollen (Fig. 79:C), ratio 0.23-0.24, propleurum with impunctate areas on lower part, vertex not lengthened or flattened (Fig. 89:A)....... ....*rotundiceps* (Provancher)

34a. (16) Abdomen, hind femur and tibia black; common in E. half of North America ...
....*praerotundiceps* sp. n.
34b. Abdomen partly orange, hind leg variable; distribution broader 35

35a. Abdominal terga 2-3 orange...36
35b. Abdominal terga 2 and sometimes 3 partly black 39

36a. Flagellum short and stout (Fig. 80:D), abdomen weakly bicolored with blackish on apical half, hind femur swollen (Fig. 79:D), its ratio 0.25, body under 5 mm long ....*brevicornis* sp. n.
36b. Flagellum longer and more slender, abdomen variable, hind femur less swollen (Fig. 79:B,C), ratio less than 0.25, body larger than 5 mm.........................37

37a. Tegula orange; Pacific Northwest.................................*xanthopyrrhus* sp. n.
37b. Tegula white; E. half of North America..38

38a. Abdomen distinctly bicolored, terga 4-apex black, hind femur moderately swollen (Fig. 79:C), clypeus wide, 3 times wider than high.....................*callistus* sp. n.
38b. Abdomen weakly bicolored, terga 5-apex blackish, hind femur slender (Fig. 79:A), clypeus broadly oval, less than 3 times wider than high...............*angularis* sp. n.

39a. Abdominal terga 2-6 partly black 40
39b. Abdominal terga 2 and sometimes 3 partly black 47

40a. Hind femur mostly black..41
40b. Hind femur orange or yellow with black apically.....................................44

41a. Clypeus black; southern California 32
41b. Clypeus white; distribution broader 42

42a. Coxae mostly white except basally, abdomen mostly black and shiny, 3-5 mm long; Arizona...*callidius* sp. n.
42b. Coxae mostly blackish except white apically, abdominal terga 2-5 mostly orangish; larger than 5 mm; western U.S..43

43a. Tegula black or brown, postpetiole elongate, 2nd abdominal tergum smooth and shiny, 5th abdominal sternum complete medially; Colorado ....*albior* sp. n.
43b. Tegula white or yellow, postpetiole nearly square, 2nd abdominal tergum slightly mat and wrinkled, 5th sternum membranous medially; northern U.S.A., Alaska, and southern Canada ....*nemati* sp. n.

44a.   Abdomen entirely black and shiny, small, under 4mm long; E. half of U.S. and
        Ontario .... *elegantulus* sp. n.
44b.   Abdomen with orange on terga 2-5, over 5 mm long; western U.S 45

45a.   Clypeus black; southern California .... *punctatior* sp. n.
45b.   Clypeus yellow or white; distribution broader 46

46a.   1st abdominal tergum slender, nearly parallel sided, postpetiole elongate (Fig.
        84:G); California .... *gracilis* sp. n.
46b.   1st abdominal tergum moderately slender, postpetiole nearly square to a little longer
        than wide (Fig. 84:F); Colorado and Montana. .... *paludicola* (Brues)

47a.   (39.) Hind femur mostly black 48
47b.   Hind femur mostly orange, apically black 49

48a.   Apophysis strong and toothlike, areola widely hexagonal (Fig. 85:D), clypeal
        margin not or weakly upturned (Fig. 78:A); Alaska to E. North America ....
        .... *pubescens* (Provancher)
48b.   Apophysis weak or absent, areola broadly hexagonal (Fig. 85:C), clypeal margin
        strongly upturned (Fig. 78:E); Arizona .... *arizonae* sp. n.

49a.   1st abdominal tergum moderately slender, postpetiole elongate (Fig. 84:G) 50
49b.   1st abdominal tergum stout, postpetiole nearly square (Fig. 84:E,F) 51

50a.   Hind femur slender (Fig. 79:A), ratio 0.19-0.20; E. of the Rocky Mts ....
        .... *chrysoleptus* sp. n.
50b.   Hind femur moderately swollen (Fig. 79:C), ratio 0.23-0.24; California ....
        .... *tricoloratus* sp. n.

51a.   Clypeus yellowish orange, hind femur orange, abdomen orange apically,
        postpetiole slightly swollen dorsally, dorsomedian carinae weak; S. and E. half of
        U.S .... *arkansensis* sp. n.
51b.   Clypeus white or yellow, hind femur black apically, abdomen black apically,
        postpetiole not swollen, dorsomedian carinae distinct; E. half of U.S. 52

52a.   Clypeal margin sharp (Fig. 78:B), face very finely granular with very fine, dense
        pits, hind femur swollen (Fig. 79:D), ratio 0.25-0.28, abdominal terga 1-2 partly
        black, propleurum entirely punctate with moderately dense pits ....
        .... *inflatus* (Provancher)
52b.   Clypeal margin not or weakly upturned (Fig. 78:A), face finely rugulose,
        punctation very dense but distinct, hind femur moderately swollen (Fig. 79:C),
        ratio 0.23-0.24, abdominal terga 1-3 partly black, propleurum impunctate on lower
        part .... *rotundiceps* (Provancher)

53a. (1.) Coxae orange or yellow...54
53b. Coxae partly or entirely black 65

54a. Abdomen, legs, scape, and clypeus mostly yellow; Mexico ....*flavissimus* sp. n.
54b. Body mostly orange and black...55

55a. 3 tyloids...56
55b. 2 tyloids...61

56a. Clypeus yellow...*xanthostomus* sp. n.
56b. Clypeus orange or black 57

57a. Clypeus not or weakly upturned (Fig. 78:A,B), apophysis strong and toothlike, mesopleurum rugulose; Pacific Northwest 58
57b. Clypeus sharply upturned (Fig. 78:D,E), apophysis weak, mesopleurum smooth; southern California and Arizona 60

58a. Abdominal terga 4-6 partly blackish, hind femur black apically ....
.... *rhyssotexanus* sp. n.
58b. Abdominal terga 5-6 orange, hind femur usually entirely orange....................59

59a. Head and thorax with orange patches, hind tibia black apically, clypeus usually orange ....*rubescens* sp. n.
59b. Head and thorax without orange patches, hind tibia entirely orange, clypeus always black...*oregonianus* sp. n.

60a. Legs and abdomen entirely orange, postpetiole elongate (Fig. 84:G); Arizona ....
.... *aureolus* sp. n.
60b. Hind femur and tibia and abdomen black apically, postpetiole square (Fig. 84:F); southern California ....*julianus* sp. n.

61a. Abdominal terga 4-6 partly or entirely black...62
61b. Abdominal terga 4-6 entirely orange...106

62a. Clypeus yellow, scape black; western U.S. and Canada ....*xanthostomus* sp. n.
62b. Clypeus orange or black, scape orange or yellow; broader distribution.............63

63a. Scape yellow, coxae and trochanters more yellowish, postpetiole wide (Fig. 84:E); N.E. half of U.S... *pseudocallistus* sp. n.
63b. Scape orange, coxae and trochanters orange or black, postpetiole wide or elongate; transcontinental.. 64

64a.   Postpetiole elongate (Fig. 84:G), clypeal margin distinctly upturned (Fig. 78:D); transcontinental...*aurantifex* sp. n.

64b.   Postpetiole a little wider than long (Fig. 84:E), clypeal margin not upturned (Fig. 78:A); transcontinental ....*rugosus* sp. n.

65a.   (53.) Tibiae mostly black with white, elongate patches dorsally, apophysis strong and toothlike, abdomen entirely black; E. North America ...*maculatus* (Provancher)

65b.   Tibiae and abdomen variable, if tibiae white dorsally, abdomen mostly orange; broader distribution.. 66

66a.   3 tyloids.. 67

66b.   2 tyloids.. 72

67a.   Clypeus yellow...*xanthostomus* sp. n.

67b.   Clypeus black 68

68a.   Scape yellow; Arizona..*tetratylus* sp. n.

68b.   Scape black or orange; Canada and northern U.S. 69

69a.   Front and middle tibiae white dorsally ....*taiganus* sp. n.

69b.   Front and middle tibiae orange or yellow dorsally...................................... 70

70a.   Abdomen 3-5 mostly orange ....*serratus* sp. n.

70b.   Abdomen 3-5 mostly black... 71

71a.   Hind femur black, hind tibia orange, flagellum short and stout (Fig. 80:D), scape orange; Alaska, Yukon, and British Columbia...................... *brachyceratus* sp. n.

71b.   Hind femur and tibia yellowish orange with black apically, flagellum moderately stout (Fig. 80:C), scape black or brown; Alaska to Greenland... *bicolor* (Lundbeck)

72a.   Abdominal terga 3-5 orange... 73

72b.   Abdominal terga 3-5 partly or entirely black.. 81

73a.   Scape yellow 74

73b.   Scape orange or black.. 75

74a.   Tegula white, apophysis weak, face and clypeus rugulose with very dense, coarse pits, clypeal margin more or less upturned (Fig. 78:C); Mexico.. *durangensis* sp. n.

74b.   Tegula black or brown, apophysis distinct, face finely granular or rugulose, clypeal margin not or weakly upturned (Fig. 78:A,B); New Mexico and Arizona............. ....*chiricahuanus* sp. n.

75a. Hind femur mostly orange or yellowish 76
75b. Hind femur mostly black... 79

76a. Tegula white, front and middle tibiae whitish dorsally; Arizona .... *auriger* sp. n.
76b. Tegula black or orange, front and middle tibiae without whitish dorsally; distribution broader... 77

77a. Postpetiole nearly square (Fig. 84:F), areola widely hexagonal (Fig. 85:D)..........
....*hexamerus* sp. n.
77b. Postpetiole elongate (Fig. 84:G), areola broadly or elongate hexagonal (Fig. 85:A,B) 78

78a. Hind femur slender (Fig. 79:A), ratio 0.18-0.20, hind femur and tibia yellowish with black apically, mesopleurum impunctate centrally .... *leioleptus* sp. n.
78b. Hind femur moderately swollen (Fig. 79:D), ratio 0.24, hind femur and tibia orange, mesopleurum rugulose and sparsely punctate centrally.. *granulifacies* sp. n.

79a. Front and middle tibiae white dorsally 80
79b. Front and middle tibiae without white dorsally..................... *pentacrocus* sp. n.

80a. Areola widely hexagonal (Fig. 85:D), mesopleurum rugulose centrally; boreal forests ....*taiganus* sp. n.
80b. Areola elongate hexagonal (Fig. 85:A), mesopleurum smooth and shiny centrally; S.W. U.S...*leucocnemis* sp. n.

81a. (72.) Abdomen entirely black 82
81b. Abdomen partly black, orange at least on terga 2 or 3............................... 88

82a. Abdominal terga 1-3 mat, mesopleurum rugulose with mostly dense, coarse pits, often sparsely punctate centrally, hind femur and tibia mostly yellowish orange; Alaska, western Canada, and Sweden .... *coriaceus* sp. n.
82b. Abdominal terga 1-2 at most slightly mat, mesopleurum mostly smooth centrally with finer punctation, hind leg black or orange; distribution broader............... 83

83a. Hind femur and scape mostly orange, abdominal terga 1-2 smooth and shiny, postpetiole elongate (Fig. 84:G) (dark form)........................*aurantifex* sp. n.
83b. Hind femur mostly black, abdominal terga 1-2 smooth or slightly mat, postpetiole variable... 84

84a.   Mesopleurum sparsely punctate centrally, hairs on head, legs, and thorax sparse
        and erect giving fuzzy appearance, scape and clypeus black; Alaska, British
        Columbia, and Scandinavia...*melanurus* (Roman)
84b.   Mesopleurum impunctate centrally, hairs not as above, more reclinate, scape and
        clypeus black or yellowish; W. North America.......................................85

85a.   Face with fine punctation, surface mostly smooth and shiny........................86
85b.   Face rugulose with coarse punctation 87

86a.   1st abdominal tergum moderately stout, postpetiole nearly square (Fig. 84:E), 2nd
        tergum regular or slightly wide, 5th abdominal sternum complete medially,
        apophysis weak...*obscurus* sp. n.
86b.   1st abdominal tergum more slender, postpetiole elongate (Fig. 84:G), 2nd tergum
        slightly narrow, 5th abdominal sternum membranous medially, apophysis distinct..
        ....*daschi* sp. n.

87a.   Apophysis weak, nearly absent, clypeus swollen, abdominal terga 1-2 slightly mat
        and wrinkled, under 4 mm long; Alaska and British Columbia.......................
        ....*minutulus* (Thomson)
87b.   Apophysis distinct, clypeus flatter, abdominal terga 1-2 usually smooth and shiny,
        sometimes with fine wrinkles, over 4 mm long; Pacific Northwest and Colorado ...
        ....*nigrans* sp. n.

88a.   (81.) Abdominal terga 1-3 mat, abdomen black on apical half, hind femur and tibia
        yellowish orange with black apically, clypeal margin distinctly upturned (Fig.
        78:D); Alaska, British Columbia, and Sweden......................... *coriaceus* sp. n.
88b.   Abdominal terga 1-2 at most mat, abdomen variable, hind femur and tibia orange or
        black, clypeal margin variable; distribution broader..................................89

89a.   Hind femur mostly orange...90
89b.   Hind femur mostly black..96

90a.   Abdomen distinctly bicolored, terga 4-6 black, postpetiole square or wide........91
90b.   Abdomen weakly bicolored, terga 4-6 mostly orange with blackish laterally and
        apcally, postpetiole elongate..94

91a.   Clypeus yellow, scape black ....*xanthostomus* sp. n.
91b.   Clypeus and scape black or orange...92

92a.   3rd abdominal tergum orange, 4th tergum partly black, front and middle trochanters orange, face rugulose with dense, coarse pits; mostly E. North America..............
....*rugosus* sp. n.

92b.   3rd abdominal tergum black apically, 4th entirely black, front and middle 1st-trochanters black, trochantelli orange, face finely rugulose or granular; mostly W. North America..93

93a.   Clypeus wide, 3 times wider than high, margin distinctly upturned (Fig. 78:D), postpetiole nearly square, flagellum moderately stout (Fig. 80:C)..... *declivis* sp. n.

93b.   Clypeus broadly oval, less than 3 times wider than high, margin not upturned (Fig. 78:A), postpetiole 1.5 times wider than long, flagellum moderately slender (Fig. 80:B) ....*latissimus* sp. n.

94a.   Postpetiole nearly square (Fig. 84:E,F), trochanters I and II orange; mostly Pacific Northwest..*hexamerus* sp. n.

94b.   Postpetiole elongate (Fig. 84:G), 1st-trochanters orange or black; distribution broader 95

95a.   Scape and legs mostly orange, hind femur moderately slender (Fig. 79:B), ratio 0.22-0.23; transcontinental...*aurantifex* sp. n.

95b.   Scape and legs mostly black, hind femur slender (Fig. 79:A), ratio 0.18-0.20; western U.S..*leioleptus* sp. n.

96a.   Tibiae I and II white dorsally, abdominal terga 2-6 orange, sometimes 1st tergum partly black ....*leucocnemis* sp. n.

96b.   Tibiae I and II without white dorsally, abdomen variable, usually black apically.. 97

97a.   Mesopleurum punctate centrally..98
97b.   Mesopleurum impunctate centrally...100

98a.   Abdominal terga 2-3 orange, postpetiole moderately slender (Fig. 84:F), areola broadly hexagonal, clypeal margin very slightly upturned (Fig. 78:C); W. North America ....*monticola* (Dalla Torre)

98b.   Abdominal terga 2-3 mostly black, postpetiole moderately slender or moderately stout, areola broadly or widely hexagonal, clypeal margin not or weakly upturned (Fig. 78:A,B); distribution broader..99

99a.   Trochantelli yellowish, flagellum moderately slender to moderately stout (Fig. 80:B,C); mostly eastern North America............................ *bicolorescens* sp. n.

99b.   Trochantelli black, flagellum stout (Fig. 80:D); Alaska, British Columbia, and Scandinavia..*melanurus* (Roman)

100a. Scape yellowish...101
100b. Scape black or orange 103

101a. Clypeus wide, 3 times wider than high, black, margin upturned (Fig. 78:C),
abdomen mostly brownish, yellowish basally on terga 3-6, coxae brownish.........
....*euryops* sp. n.
101b. Clypeus broadly oval, less than 3 times wider than high, black or yellow, margin
not or weakly upturned (Fig. 78:A,B), abdomen mostly orangish or blackish,
coxae variable 102

102a. Abdominal terga 2-3 mostly blackish, clypeus usually black, areola broadly
hexagonal (Fig. 85:C); Alaska to Oregon................................*obscurus* sp. n.
102b. Abdominal terga 2-3 mostly orange, clypeus usually yellowish orange, areola
widely hexagonal (Fig. 85:D); E. North America.................*michiganensis* sp. n.

103a. Clypeus and front and middle coxae orange, front and middle trochanters orange,
abdomen weakly bicolored, sometimes black spots on 2nd tergum; N. North
America ....*leopardus* sp. n.
103b. Clypeus and front and middle coxae black, front and middle trochanters and
abdomen variable 104

104a. Face rugulose with dense, coarse pits, propodeal carinae strong and coarse,
postpetiole and 2nd tergum wide, latter slightly mat.....................*rugosus* sp. n.
104b. Face finely rugulose or granular, smooth, propodeal carinae distinct but fine,
postpetiole a little longer than wide, 2nd tergum smooth and shiny.................105

105a. Abdominal terga 2-5 mostly orange, trochantelli orange, areola broadly hexagonal
(Fig.   85:C)... *pentacrocus* sp. n.
105b. Abdominal terga 2-5 mostly blackish, trochantelli pale yellow, areola elongate
hexagonal (Fig. 85:B)...*daschi* sp. n.

106a. (61.) Head rugulose with coarse, dense pits, clypeus blackish.....................107
106b. Head smoother, punctation dense but finer, clypeus variable, white, orange, or
black...109

107a. Apophysis strong and toothlike, postpetiole elongate (Fig. 84:G); southeastern
U.S...*spicus* sp. n.
107b. Apophysis weak, postpetiole nearly square 108

108a. Vertex flattened and slightly lengthened (Fig. 89:B), hind femur swollen (Fig. 79:D), ratio 0.25, clypeal margin impressed; Arizona and Mexico....*occipitis* sp. n.
108b. Vertex more rounded, not lengthened (Fig. 89:A), hind femur moderately slender (Fig. 79:C), ratio 0.22-0.24, clypeus more convex; E. North America................
.... *rugiceps* sp. n.

109a. Clypeus white, scape yellow, apophysis strong and toothlike; Alaska................
....*leptotexanus* sp. n.
109b. Clypeus orange or black, scape orange or black, apophysis reduced or moderately strong, not toothlike; Pacific Northwest or southwestern U.S.....................110

110a. Clypeal margin not or weakly upturned (Fig. 78:A,B), areola broadly hexagonal (Fig. 85:C); Pacific Northwest.....................................*hesperus* sp. n.
110b. Clypeal margin upturned (Fig. 78:E), areola elongate hexagonal (Fig. 85:B); Arizona..111

111a. Tegula white, scape and clypeus usually orange, postpetiole nearly square, 2nd tergum slightly wide.......................................*auriger* sp. n.
111b. Tegula orange, scape and clypeus orangish black or black, postpetiole elongate, 2nd tergum narrow ....*aureolus* sp. n.

Females
(Females of *melanogaster, flavissimus, julianus,* and *spinissimus* unknown)

1a. 3rd valvula 1.25-1.5 times longer than hind tarsomeres...............................2
1b. 3rd valvula as long as or shorter than hind tarsomeres...............................3

2a. Abdomen bicolored, on apical half, legs black; N. North America.. *coriaceus* sp. n.
2b. Abdomen and legs orange; E. North America......................*tyloidiphorus* sp. n.

3a. Flagellar segments 4-9 partly or entirely white dorsally...............................66
3b. Flagellar segments 4-9 pale yellow, orange, or black...................................4

4a. 3rd valvula as long as or shorter than basal 3 hind tarsomeres.......................5
4b. 3rd valvula longer than basal 3 hind tarsomeres 30

5a. Radius curved beyond areolet (Fig. 82:D)...6
5b. Radius mostly straight beyond areolet (Fig. 82:B,C) 9

6a.   Flagellum bicolored dorsally, areola longer than wide, abdominal terga 2-6 partly
       black; southern California...*flavivittatus* sp. n.
6b.   Flagellum orange or black dorsally, areola variable, usually shorter (Fig. 85:A),
       abdomen variable; distribution broader..7

7a.   Flagellum narrowed apical half (Fig. 80:F), basal 3 flagellomeres moniliform;
       California or transcontinental...8
7b.   Flagellum nearly linear to apex, basal 3 segments short (Fig. 80:G); California......
       ....*gracilis* sp. n.

8a.   Legs mostly orange, hind femur and tibia often black apically, face rugulose;
       transcontinental across southern Canada and northern U.S.A ....*aurantifex* sp. n.
8b.   Front and middle legs yellow, hind femur blackish, hind coxa and tibia orange, face
       finely granular; southern California................................ *santacruzensis* sp. n.

9a.   Flagellum entirely orangish or blackish dorsally......................................10
9b.   Flagellum bicolored or tricolored 16

10a.   Flagellum linear to apex (Fig. 80:F), basal 3 flagellomeres short (Fig. 80:H), radius
       beyond areolet straight (82:B)...*auriger* sp. n.
10b.   Flagellum narrowed apical half (Fig. 80:E), without above combination of
       characters 11

11a.   Cheeks conspicuously swollen, clypeus impressed, margin distinctly upturned
       (Fig. 78:D), flagellum black; central Mexico.......................*pinidiprionis* sp. n.
11b.   Cheeks normally convex, clypeus variable, flagellum usually orange; W. North
       America and Arizona 12

12a.   Coxae partly black...13
12b.   Coxae entirely orange 14

13a.   Clypeus, tegula, coxae, and trochanters black ....*daschi* sp. n.
13b.   Clypeus, tegula, most of coxae and trochanters orange.......*monticola* (Dalla Torre)

14a.   Apophysis weak and indistinct, clypeal margin strongly upturned (Fig. 78:E);
       Arizona...*tetratylus* sp. n.
14b.   Apophysis strong and distinct, clypeal margin not or weakly upturned (Fig.
       78:A,B); W. and N. North America..15

15a. Hind femur and tibia orange with black apically, hind femur slender (Fig. 79:B,C), ratio 0.23-0.25 .... *rhyssotexanus* sp. n.

15b. Hind femur and tibia entirely orange, hind femur moderately swollen (Fig. 79:D), ratio 0.25-0.28 79

16a. Flagellum linear to apex 17

16b. Flagellum narrowed apical half 23

17a. Coxae I and II yellow, coxa III orange or yellow 18

17b. Coxae black or brownish 20

18a. Clypeal margin upturned (Fig. 78:C,D), clypeus black or blackish orange; Mexico or southeastern U.S. 19

18b. Clypeal margin not or weakly upturned (Fig. 78:A,B), clypeus orange; E. North America .... *rotundiceps* (Provancher)

19a. Basal 3 flagellomeres elongate (Fig. 80:I), clypeus black, temple rugulose with coarse pits; Mexico .... *durangensis* sp. n.

19b. Basal 3 flagellomeres short (Fig. 80:H), clypeus blackish orange, temple not rugulose, punctation variably sparse to dense with finer pits; southeastern U.S .... *arkansensis* sp. n.

20a. Abdomen orange, clypeal margin distinctly upturned (Fig. 78:D); Arizona .... *concavus* sp. n.

20b. Abdomen partly black, clypeal margin not or weakly upturned (Fig. 78:A-C) 21

21a. Clypeus wide, about 3 times wider than long, margin upturned (Fig. 78:C); transcontinental .... *euryops* sp. n.

21b. Clypeus broadly oval, less than 3 times wider than long, margin not or weakly upturned (Fig. 78:A,B) 22

22a. Abdomen bicolored, terga 2-3 orange, remaining terga black, scape orange .... *pentacrocus* sp. n.

22b. Abdominal terga 2-6 blackish, scape black .... *obscurus* sp. n.

23a. Coxae partly or entirely black 24

23b. Coxae orange or yellow 26

24a. Clypeal margin strongly upturned (Fig. 78:D,E); Oregon and California .... *granulifacies* sp. n.

24b. Clypeal margin not or weakly upturned (Fig. 78:A,B); Alaska and British Columbia, or Colorado and Montana 25

25a. Coxae and femora black, abdomen distinctly bicolored, apophysis weak; Alaska,
     British Columbia, and Scandinavia...*melanurus* (Roman)
25b. Coxae and femora more yellowish or orange, abdomen weakly bicolored,
     apophysis distinct; Colorado and Montana. .... *paludicola* (Brues)

26a. Clypeus impressed, margin upturned (Fig. 78:C), black, scape yellowish orange;
     Arizona and Mexico ....*occipitis* sp. n.
26b. Clypeus convex, margin not or weakly upturned (Fig. 78:A,B), mostly orange,
     scape orange or black; western or eastern U.S..27

27a. Hind femur moderately slender (Fig. 79:B,C), ratio 0.22, 1st abdominal tergum
     moderately slender, postpetiole about 1.5 times wider than long (Fig. 84:C),
     apophysis strong and toothlike; Alaska ....*leptotexanus* sp. n.
27b. Hind femur moderately swollen (Fig. 79:D), ratio 0.24 or greater, 1st abdominal
     tergum stout, postpetiole about twice as wide as long (Fig. 78:A,B), apophysis
     distinct but not toothlike; distribution other than Alaska 28

28a. Hind tibia and sometimes femur black apically, 3rd valvula about as long as basal 2
     hind tarsomeres or shorter; E. North America........................*patulus* (Viereck)
28b. Hind tibia and hind femur orange, 3rd valvula about as long as basal 3 hind
     tarsomeres; W. North America...29

29a. Flagellum tricolored dorsally, segments 5-9 yellowish orange, postpetiole width
     greater than twice the length..*rubescens* sp. n.
29b. Flagellum bicolored dorsally, segments 1-9 uniformly orange, postpetiole about
     1.75 times as wide as long...*hexamerus* sp. n.

30a. (4) Flagellum bicolored dorsally 31
30b. Flagellum orange or black 54

31a. Flagellum narrow apical half (Fig. 80:E) 32
31b. Flagellum linear to apex (Fig. 80:F)...37

32a. Clypeal margin distinctly upturned (Fig. 78:D,E) 33
32b. Clypeal margin not or weakly upturned (Fig. 78:A,B)................................34

33a. Front and middle legs yellow, frons very densely pitted; eastern U.S. and Ontario..
     .... *rugiceps* sp. n.
33b. Front and middle legs orange, frons sparsely pitted; Arizona .... *aureolus* sp. n.

34a. Clypeus orange, hind femur slender (Fig. 79:B), ratio 0.23 ....*xanthostomus* sp. n.
34b. Clypeus black, hind femur more swollen (Fig. 79:C,D), ratio 0.24-0.28..........35

35a. Abdominal terga 3-6 orange; Minnesota to New York . . . . *pseudocallistus* sp. n.
35b. Abdominal terga 3-6 partly or entirely black; W. North America 36

36a. Coxae black; Alaska, Yukon, and British Columbia . . . . *brachyceratus* sp. n.
36b. Coxae black with yellowish apically; Colorado and Montana . . . . . *paludicola* (Brues)

37a. Abdominal terga 2-6 orange 38
37b. Abdominal terga 2-6 partly or entirely black 44

38a. Hind coxa blackish, hind femur and tibia orange; Arizona and New Mexico . . . .
      . . . . *chiricahuanus* sp. n.
38b. Hind coxa usually orange, if blackish, hind femur and tibia black apically; E. North
      America 39

39a. Thorax and propodeum orange . . . . *aurarius* sp. n.
39b. Thorax and propodeum black 40

40a. Clypeal margin distinctly upturned (Fig. 78:D), flagellum slightly tricolored
      dorsally, segments 5-9 pale yellow 41
40b. Clypeal margin not or weakly upturned (Fig. 78:A,B), flagellum usually bicolored
      dorsally 42

41a. Hind femur swollen (Fig. 79:E), ratio 0.29-0.30, vertex slightly lengthened and
      flattened (Fig. 89:B), apophysis about as long as wide, areola broadly hexagonal
      (Fig. 85:C), face finely rugulose with very dense, fine pits . . . . *inflatus* (Provancher)
41b. Hind femur moderately swollen (Fig. 79:D), ratio 0.25, vertex not so lengthened or
      flattened (Fig. 89:A), apophysis toothlike, longer than wide, areola elongate
      hexagonal (Fig. 85:A), a little arched, face rugulose with dense, coarse pits . . . .
      . . . . *auriculiferus* (Viereck)

42a. Basal 3 flagellomeres elongate (Fig. 80:I), hind femur slender (Fig. 79:B), ratio
      0.21, hind coxa often blackish, 3rd valvula about as long as hind tarsomeres,
      clypeus black . . . . *chrysoleptus* sp. n.
42b. Basal 3 flagellomeres short (Fig. 80:H), hind femur moderately swollen (Fig.
      79:D), ratio 0.24 or greater, hind coxa orange, 3rd valvula about as long as basal 4
      hind tarsomeres, clypeus black or orange 43

43a. Temple rugulose with dense to moderately dense, more or less distinct pits (Fig.
      86:D), frons rugulose with very dense, coarse pits, hind femur and tibia usually
      entirely orange . . . . *subclavatus* (Say)
43b. Temple smooth with moderately dense, indistinct pits (Fig. 86:C), frons slightly
      rugulose with dense, fine pits, hind femur and tibia black apically . . . *aurigena* sp. n.

44a. (37) Abdomen entirely black, body under 4mm long 45
44b. Abdomen partly orange or yellow, body size variable 46

45a. Coxae and trochanters yellow; E. North America.................... *elegantulus* sp. n.
45b. Coxae and trochanters black; Alaska, British Columbia, and Europe ....
            .... *minutulus* (Thomson)

46a. Coxae I and II yellow 47
46b. Coxae I and II black, brownish, or blackish orange.................................. 49

47a. Abdominal terga 1-3 black, remaining terga more yellowish, 1st tergum moderately
      slender (Fig. 84:C) .... *elegantulus* sp. n.
47b. Abdominal terga 1-3 more yellow or orange, remaining terga more blackish, 1st
      tergum moderately stout (Fig. 84:B) 48

48a. Abdomen weakly bicolored, terga 2-3 mostly orange, hind coxa yellow with black
      basally, hind tibia mostly yellowish black, 1st abdominal tergum abruptly widened
            .... *brevicornis* sp. n.
48b. Abdomen not bicolored, terga 2-3 yellow and blackish, hind coxa yellow or
      blackish yellow, 1st tergum gradually widened, hind tibia yellow with black
      apically .... *praerotundiceps* sp. n

49a. Hind femur and tibia yellowish orange, apophysis weak or absent; N. North
      America and Greenland .... *bicolor* (Lundbeck)
49b. Hind femur partly or entirely black, hind tibia partly black, apophysis distinct;
      Alaska or mostly W. and E. North America... 50

50a. Clypeal margin upturned (Fig. 78:C), wide—3 times wider than high, abdomen
      and hind femur brownish .... *euryops* sp. n.
50b. Clypeal margin not or weakly upturned (Fig. 78:A,B), broadly oval, less than 3
      times wider than high, abdomen orange and black, hind femur black.............. 51

51a. Front and middle coxae yellow with black basally, abdomen mostly orange except
      laterally.. *pubescens* (Provancher)
51b. Front and middle coxae black or blackish orange, abdomen mostly black or black
      apically (bicolored) 52

52a. Hind coxa orange, radius curved beyond areolet (Fig. 82:D), front and middle
      coxae blackish orange .... *michiganensis* sp. n.
52b. Hind coxa mostly black, radius mostly straight beyond areolet (Fig. 82:B,C), front
      and middle coxae black... 53

53a. Scape and tegula black, basal 3 flagellomeres short (Fig. 80:H)......*obscurus* sp. n.
53b. Scape and tegula yellowish, basal 3 flagellomeres elongate (Fig. 80:I)...............
.... *bicolorescens* sp. n.

54a. (30.) Flagellum linear to apex (Fig. 80:F) 55
54b. Flagellum narrow apical half (Fig. 80:E) 60

55a. Abdominal terga 2-6 orange, hind leg orange; southern California....................
....*punctatior* sp. n.
55b. Abdominal terga 2-6 partly black, hind leg partly black; W. North America and
Arizona.. 56

56a. Hind coxa orange, front and middle coxae yellow; southern California ....
.... *tricoloratus* sp. n.
56b. Hind coxae partly or entirely black, front and middle coxae variable; W. North
America or Arizona 57

57a. Coxae I and II yellow, clypeus and tegula yellow; Arizona ....*callidius* sp. n.
57b. Coxae I and II mostly black or orange, clypeus usually black or orange, tegula
variable; W. North America 58

58a. Front and middle coxae mostly black, trochanters yellow, face rugulose with dense,
coarse pits, radius curved beyond areolet (Fig. 82:D)....................*nigrans* sp. n.
58b. Front and middle coxae mostly orange, trochanters orangish, face finely rugulose
or granular with very dense, fine pits, radius straight beyond areolet (Fig.
82:B,C).. 59

59a. Abdomen weakly bicolored, black on terga 3-6 apically, 1st tergum stout and wide,
postpetiole twice as wide as long, radial cell distinctly longer than 2nd discoidal cell
....*leopardus* sp. n.
59b. Each abdominal tergum partly or entirely black, 1st tergum moderately stout (Fig.
84:B), postpetiole less than twice as wide as long, radial cell about as long as 2nd
discoidal cell..*nemati* sp. n.

60a. (54) Abdomen mostly orange 61
60b. Abdomen distinctly bicolored 63

61a. Clypeal margin strongly upturned (Fig. 78:E); Arizona.................*arizonae* sp. n.
61b. Clypeal margin not or weakly upturned (Fig. 78:A,B); Pacific Northwest......... 62

62a. Clypeus black, frons rugulose with very dense, coarse pits.......*oregonianus* sp. n.
62b. Clypeus orange, frons finely rugulose with very dense, fine pits ....
.... *xanthopyrrhus* sp. n.

63a.   Coxae black.. 64
63b.   Coxae partly orange or yellow.. 65

64a.   Clypeus wide, 3 times wider than long, flagellum black dorsally.......*declivis* sp. n.
64b.   Clypeus broadly oval, less than 3 times as wide as long, flagellum orange dorsally
        ....*leioleptus* sp. n.

65a.   Hind femur swollen (Fig. 79:E), ratio 0.29, basal 3 flagellomeres short (Fig.
        80:H)..*serratus* sp. n.
65b.   Hind femur moderately swollen (Fig. 79:D), ratio 0.25, basal 3 flagellomeres
        elongate (Fig. 80:I).. *albior* sp. n.

66a.   (3) Flagellum linear to apex (Fig. 80:F), clypeus distinctly upturned (Fig. 78:D).....
        ....*spicus* sp. n.
66b.   Flagellum narrow apical half (Fig. 80:E), clypeal margin not or weakly upturned
        (Fig. 78:A,C).. 67

67a.   Tibiae black with white elliptical patches dorsally, abdomen entirely black.............
        ....*maculatus* (Provancher)
67b.   Tibiae variable, if front and middle tibiae white dorsally, not elliptical patches,
        abdomen never entirely black...68

68a.   Abdomen bicolored.. 69
68b.   Abdomen orange or blackish orange.. 73

69a.   Coxae orange.. 70
69b.   Coxae partly or entirely black...72

70a.   Abdomen weakly bicolored, terga 6-apex black...........................*rugitexanus* sp. n.
70b.   Abdomen distinctly bicolored, terga 4-apex black... 71

71a.   Clypeus wide, 3 times wider than long, margin upturned (Fig. 78:C); N. and E.
        half of U.S.A..*callistus* sp. n.
71b.   Clypeus broadly oval, less than 3 times wider than long, margin not upturned (Fig.
        80:A); Alaska, British Columbia, and Newfoundland.....................*latissimus* sp. n.

72a.   Tegula and coxae black, hind femur and tibia orange, clypeus wide, 3 times wider
        than long.. *declivis* sp. n.
72b.   Tegula orange, coxae blackish orange, hind femur blackish, hind tibia orange with
        black apically, clypeus broadly oval, less than 3 times wider than long...................
        ....*rugosus* sp. n.

73a. Coxae pale yellow ventrally..*rotundiceps* (Provancher)
73b. Coxae black or orange ventrally...74

74a. Coxae orange..75
74b. Coxae black..77

75a. Hind tarsus whitish..*albitexanus* sp. n.
75b. Hind tarsus orange or black..76

76a. Clypeus and mandible orange, 3rd valvula about as long as basal 2 hind tarsomeres, hind leg orange..*angularis* sp. n.
76b. Clypeus and mandible black, 3rd valvula about as long as basal 4 hind tarsomeres, hind femur black apically, hind tarsomeres mostly black.............*texanus* (Cresson)

77a. Scape orange, hind tibia black, usually with white stripe dorsally, trochantelli whitish; E. N. America.....................................*mucronatus* (Provancher)
77b. Scape black, hind tibia orangish or black without white dorsally, trochantelli black; southwestern U.S. or mostly W. North America......................................78

78a. Hind tibia orange with black apically, areola widely hexagonal (Fig. 85:D); mostly Alaska, Yukon, and British Columbia................................... *taiganus* sp. n.
78b. Hind tibia black, areola broadly hexagonal (Fig. 85:C); Arizona, New Mexico, and western Texas... *leucocnemis* sp. n.

79a. (15) Face finely granular or finely rugulose, wings more or less darkened, femoral ratio about 0.28, flagellum slender, basal 3 segments slightly elongate...................
....*hesperus* sp. n.
79b. Face rugulose with very dense pits, wings distinctly blackish, femoral ratio about 0.26, flagellum moderately slender, basal segment long, 2nd and 3rd short.............
....*hexamerus* sp. n.

TEXANUS GROUP

DIAGNOSIS. Clypeal margin not or weakly upturned; propodeum more or less level in side view, carinae strong and often coarse, apophysis strong and toothlike; dorsomedian carina of 1st abdominal tergum strong beyond spiracle; tibiae often with white dorsally; males usually with 3 tyloids on flagellar segments 10-12; female often with white on flagellar segments 5-9, basal 3 segments elongate.

This is the most morphologically generalized group of species. I have included 16 Nearctic species, although *rugosus* and *pseudocallistus* are included with reservations. The group can be characterized as follows: clypeus convex, margin not or only very slightly upturned; face width about equal to height of face plus clypeus; female flagellum narrowed apically, basal 3 segments distinctly elongate, and often with distinct annulus on flagellar segments 4 or 5 to 9 or 10; male flagellum often with 3rd tyloid on 12th segment; punctation of head evenly dense with distinct pits, surface more or less rugulose, male mesopleural plate usually rugulose centrally; males with white typically on front and middle tibiae dorsally, and often on coxae apically, trochanters ventrally, tegula, or scape; females often with white dorsally on front and middle tibiae, and sometimes on scape or trochanters; propodeum more or less level in lateral view along lateral carina, carinae coarse and strong, apophysis strong and toothlike, area dentipara distinctly narrowed, often nearly triangular; male 1st abdominal tergum moderately stout to moderately slender, postpetiole squarish, dorsomedian carina strong beyond spiracle; female 1st tergum moderately slender, gradually widened to about 1.5 times length, dorsomedian carina distinct; radial cell of both male and female moderately long, Rs straight and very slightly bowed, areolet more or less narrow; hind femur moderately slender to slender; males with 5th abdominal sternum often completely sclerotized, 6th always, at least apically; dimorphism less marked; mostly large species.

The species in this group are most common in the Pacific Northwest, Alaska to British Columbia, and the eastern half of the United States and southern Canada. One species occurs in southern California, but not in Arizona and Mexico. Recorded hosts are argid sawflies which feed on birch, willow, alder, and several other genera of non-gymnospermous trees of boreal forests.

This group corresponds to the Palearctic Cnemargus Group—including 4 species The following Nearctic species are included: *albitexanus, angularis, latissimus, leptotexanus, maculatus* (Provancher), *mucronatus* (Prov.), *oregonianus*, *pseudocallistus*, *rhyssotexanus*, *rubescens*, *rugitexanus*, *rugosus*, *taiganus*, *texanus* (Cresson), *xanthopyrrhus*, and *xanthostomus*.

Endasys albitexanus Luhman, sp. n.
(Fig. 69: A-D; Map 2)

MALE DIAGNOSIS. Large, 7.5-8.5 mm long; white hind tarsomeres, clypeus, scape, coxae I and II, and coxa III apically; femur III and tibia III orange except both dorso-apically black; areola widely hexagonal, nearly quadrangular, area dentipara narrowed, apophysis strong and toothlike, projecting vertically; 3rd tyloid on flagellar segment 12; 1st and 2nd terga smooth and shiny; 5th abdominal sternum membranous except apically; eastern half of United States and southern Canada.

MALE DESCRIPTION. *White*: Scape, clypeus, mandible, tegula, coxae I and II except basally, coxa III apically, trochanters, all tarsomeres. *Yellow*: Flagellum more yellowish ventrally, coxae I and II basally, femora I and II, tibiae. *Orange*: Coxa III except apically, femur III and tibia III except dorso-apically, abdomen except parameres. *Black*: Flagellum except ventrally, femur III and tibia III dorso-apically, parameres. *Punctation* (Fig. 86:D-E): Face and frons finely rugulose with very dense pits; temple with dense, more or less distinct pits; propleurum with dense to sparse, distinct pits, surface variably rugulose; mesopleurum impunctate centrally and rugulose, otherwise with dense to sparse, distinct pits, surface with rugulosity; 1st and 2nd terga smooth and shiny. *Shape*: Clypeal margin sharp but not distinctly upturned (Fig. 78:B); flagellum moderately slender (Fig. 80:B), its segments longer than wide to apex; areola widely hexagonal, nearly quadrangular and somewhat linear, area dentipara narrowed, apophysis strong and toothlike, vertically projecting, propodeal carinae distinct; 1st tergum moderately slender (Fig. 84:F), postpetiole square, 2nd tergum a little widened; femur III moderately swollen (Fig. 79:C), ratio 0.22; radial cell long, 1.5 times longer than 2nd-discoidal cell, Rs long. *Other*: Short, 3rd tyloid on flagellar segment 12; wings slightly tinted; 5th abdominal sternum membranous medially, but complete apically (Fig. 81:B).

FEMALE DIAGNOSIS. Large, 8-9.5 mm long; hind tarsomeres yellowish; 1st abdominal tergum moderately slender, postpetiole abruptly widened; apophysis very strong and toothlike, vertically projecting; flagellum moderately slender, narrowed apically (Fig. 80:E), flagellar segment 4 or 5 to 9 white.

FEMALE DESCRIPTION. *White*: Flagellar segments 4 or 5-9. *Yellow*: Legs I and II, hind trochanters and tarsomeres. *Orange*: Scape often, mandible, tegula, coxa III, femur III and tibia III except dorso-apically, abdomen. *Black*: Flagellum except annulus, scape sometimes more blackish, clypeus, femur III and tibia III dorso-apically. *Punctation* (Fig. 86:F): Face, frons, propleurum, and mesopleurum rugulose with very dense, coarse pits; temple variably rugulose with dense to very dense, coarse pits. *Shape*: Clypeus a little impressed apically, margin sharp but only very slightly upturned (Fig. 78:B); flagellum moderately slender, narrowed apically, basal 3 segments distinctly elongate (Fig. 80:E,I);

areola widely hexagonal or broader, area dentipara narrowed, elongate apically, apophysis strong and toothlike, vertically projecting, propodeal carinae weak; femur III swollen (Fig. 79:E), ratio 0.26-0.27; 1st tergum moderately slender, postpetiole abruptly widened to more than 1.5 times the length; Rs straight, radial cell long, 1.3 times the length of the 2nd-discoidal cell. *Other*: 3rd valvula short or moderately short, about as long as basal 3 hind tarsomeres; wings slightly tinted.

REMARKS. See discussion under *rugitexanus*.

RANGE. Minnesota, Michigan, and Ontario southeastward to New York, West Virginia, South Carolina, and Georgia. Collections May, June, and September.

MATERIAL EXAMINED. **Holotype**: male, Bowden, WEST VIRGINIA, 6-7-vi-1980, C. Dasch (AEI). **Paratypes**: 2 M, 5 F (AEI, UGA): GEORGIA, MICHIGAN, MINNESOTA, NEW YORK, ONTARIO, SOUTH CAROLINA.

ETYMOLOGY. *Albi-* (white) + *texanus*, referring to the whitish hind tarsomeres of this Texanus Group species.

Endasys angularis Luhman, sp. n.
(Fig. 57:A-D; Map 1)

MALE DIAGNOSIS. Medium large, about 6.5 mm long; propodeum coarsely rugulose with strong carinae; 1st and 2nd abdominal terga widened (Fig. 84:D) and strongly mat; femur III slender (Fig. 79:A), ratio 0.19; flagellum long and moderately slender (80:B), tyloids (2) long and narrow; clypeus white, margin not upturned; white coxae I and II and all trochanters, yellow scape, orange femur III, orange tibia III with blackish apex, abdomen weakly bicolored with black on at least terga 5 to apex; northern United States.

MALE DESCRIPTION. *White*: Clypeus, mandible, tegula, coxae I and II except basally, trochanters. *Yellow*: Scape, front and middle coxae basally, femora I and II, tibiae I and II. *Orange*: Flagellum pale orange ventrally, coxa III, femur III except often apically, tibia III except apex, hind tarsomeres orangish, abdominal terga 1 to 4 or 5, at least basally. *Black*: Flagellum dorsally, often femur III apically, tibia III apically, terga 6 to apex, at least laterally and apically on terga 4-5. *Punctation* (Fig. 86:C-D,F): Face and frons finely rugulose with very dense pits; temple with variably dense to sparse, indistinct pits; propleurum with dense to sparse, more or less distinct pits, surface a little rugulose; mesopleurum impunctate centrally, otherwise rugulose with more or less distinct pits; 1st and 2nd terga strongly mat and slightly wrinkled. *Shape*: Clypeal margin not upturned (Fig. 78:A); flagellum long and moderately slender (Fig. 80:B), its segments longer than wide to apex; areola widely hexagonal, nearly quadrangular, area dentipara strongly narrowed, apophysis distinct, propodeum short and rugulose with strong carinae; 1st tergum stout (Fig. 84:D), postpetiole wider than long, spiracle close to midlength, 2nd tergum distinctly wide; femur III slender (Fig. 79:A), ratio 0.19; radial cell long, Rs straight, areolet wider than long. *Other*: Tyloids (2) long and narrow (Fig. 87:A), not prominent; wings slightly tinted; 5th abdominal sternum membranous medially.

FEMALE DIAGNOSIS. Medium large, 6-6.5 mm long; white flagellar segments 4-9, legs and abdomen orange, 3rd valvula short, as long as basal 2 hind tarsomeres, femur III slender (Fig. 79:B), ratio 0.22-0.23, 1st abdominal tergum stout and widened (Fig. 84:A), propodeum short and angular, area dentipara distinctly narrowed, apophysis distinct.

FEMALE DESCRIPTION. *White*: Flagellar segments 4 or 5 to 9. *Orange*: Flagellar segments 1 to 3 or 4, scape, clypeus, mandible, tegula, legs, and abdomen except laterally on terga 4-7. *Black*: Flagellar segments 10-apex, sometimes femur III blackish apically, terga 4 to 7 laterally blackish. *Punctation* (Fig. 86:B-D): Face finely rugulose with very dense pits; frons with dense, very fine, distinct pits; temple with dense hairs, pits indistinct; propleurum with dense, more or less distinct pits; mesopleurum with variably sparse to dense punctation, more or less distinct pits, surface with rugulosity. *Shape*: Clypeal margin not upturned (Fig. 78:A); flagellum moderately stout, narrowed apically, basal 3 segments elongate (Fig. 80:E,I); areola widely hexagonal, area dentipara strongly narrowed, elongate apically, apophysis distinct and wide, propodeum short, carinae distinct; 1st tergum stout and wide (Fig. 84:A), twice as wide as long; femur III slender (Fig. 79:B), ratio 0.22-0.23; radial cell moderately long, Rs slightly curved. *Other*: 3rd valvula short, as long as basal 2 hind tarsomeres; wings a little tinted.

REMARKS. See discussion under *rugitexanus*.

RANGE. Michigan and New York. Collection May and June.

MATERIAL EXAMINED. **Holotype**: male, Ann Arbor, MICHIGAN, 31-vi-1959, H. and M. Townes (AEI). **Paratypes**: 7 M, 4 F (AEI, AMNH, Dasch): MICHIGAN, NEW YORK. Other material studied: 1 F (AEI): MICHIGAN.

ETYMOLOGY. *Angularis* (angular), referring to the coarse, angular propodeum.

Endasys latissimus Luhman, sp. n.
(Fig. 26:A-D; Map 5)

MALE DIAGNOSIS. Medium large, 5.5-7 mm long; black coxae, 1st trochanters, femur III and tibia III apically, abdomen basally and terga 4-apex, clypeus and mandible, and flagellum and scape; remainder of legs and abdomen orange; propodeum short, sloped apically, area dentipara distinctly narrowed and elongate, apophysis distinct; 1st abdominal tergum stout (Fig. 84:D) and wide, postpetiole moderately widened, about 1.5 times as wide as long, 2nd tergum wide, terga 1 and 2 mat; clypeal margin not upturned; punctation of face and frons generally very dense, surfaces granular or slightly rugulose, temple sparsely and finely pitted; Alaska, western Canada, and Colorado.

MALE DESCRIPTION. *Orange*: Trochantelli, femora except femur II apically, femur III ventrally, tibiae except tibiae I and II apically, usually 1st abdominal tergum apically, and terga 2-4 except 4 apically. *Black*: Flagellum, scape, clypeus, mandible, tegula, coxae, 1st trochanters, femur III apically, tibia III apically, hind tarsomeres, 1st tergum often entirely, terga 4-apex except 4 basally. *Punctation* (Fig. 86:A-B,D-E): Face granular with very dense pits; frons slightly rugulose with variably dense to very dense, very fine pits; temple with moderately sparse, fine, indistinct pits; propleurum with dense to very sparse, fine, more or less indistinct pits, surface shiny and smooth; mesopleurum variably sparse to very sparse centrally with some rugulosity, otherwise rugulose with mostly sparse, distinct pits; 1st and 2nd terga distinctly mat. *Shape*: Clypeal margin not upturned (Fig. 78:A); flagellum moderately slender (Fig. 80:B), its segments longer than wide to apex; areola widely hexagonal, area dentipara distinctly narrowed and a little elongate, apophysis distinct, propodeal carinae distinct, propodeum short and wide, sloped apically; 1st tergum stout and wide (Fig. 84:D), postpetiole about 1.5 times as wide as long, 2nd tergum wide; femur III moderately slender (Fig. 79:B), *Other*: Wings a little tinted, 5th abdominal sternum membranous medially, glumes dense (Fig. 88:A), and tyloids (2) short.

FEMALE DIAGNOSIS. Medium large, 5.5-7 mm long; flagellum white on segments 5 to 8 or 9, moderately swollen medially, narrowed apically (Fig. 80:E), basal 3 segments elongate; 3rd valvula long, about as long as hind tarsomeres; femur III and tibia III orange with black apices, legs I and II orange; 1st abdominal terga stout and wide (Fig. 84:A), postpetiole about twice as wide as long; propodeum shortened and a little sloped apically, area dentipara distinctly narrowed, a little elongate, apophysis distinct, projecting at an angle; abdomen bicolored, orange on terga 1-3 and 4 basally, and black on terga 4 apically to apex; femur III moderately swollen (Fig. 79:C), ratio about 0.25.

FEMALE DESCRIPTION. *White*: flagellar segments 5 to 8 or 9. *Orange*: Scape, mandible, tegula, legs except both femur III and tibia III apically, and abdomen except terga 4-6 apically and apex. *Black*: Flagellar segments 1-4 and 9 or 10 to apex, clypeus more orangish black, femur III apically, tibia III apically, hind tarsomeres, abdominal terga 4-6 apically, apex. *Punctation* (Fig. 86:C-E): Face very finely granular with dense to very dense pits, surface a little shiny; frons with dense to very dense, distinct pits; temple with moderately sparse, indistinct pits; propleurum with dense to sparse, more or less distinct pits, mostly smooth and shiny; mesopleurum with mostly sparse, more or less distinct pits, surface very slightly rugulose. *Shape*: Clypeal margin not upturned (Fig. 78:A); flagellum moderately swollen medially but narrowed apically, basal 3 segments elongate (Fig. 80:E,I); areola broadly hexagonal, area dentipara narrowed, elongate apically, apophysis distinct, propodeum short and a little sloped apically; 1st tergum stout and wide (Fig. 84:A), postpetiole about twice as long as wide; femur III moderately swollen (Fig. 79:C), ratio 0.25; radial cell long, Rs straight (Fig. 82:B). *Other*: 3rd valvula long, about as long as hind tarsus; wings blackish.

REMARKS. This species is related to *pseudocallistus,* males of which are distinguished by the white scape and yellowish orange coxae I and II; females differ from *latissimus* by the bicolored flagellum, without white annulus on segments 5-9. *Latissimus* appears similar to Palearactic *analis* (Thomson) which, however, has a distinctly swollen face and very weakly bicolored abdomen. If they are related, then *latissimus* should be in the Monticola Group.

RANGE. Alaska, western Canada, Newfoundland, and Colorado. Collections June through August.

MATERIAL EXAMINED. **Holotype**: male, Stone Mt. Pk., 5500 ft., BRITISH COLUMBIA, 18-vii-1973,   H. and M. Townes (AEI). **Paratypes**: 19 M, 2 F (AEI, CNC, Dasch, USNM): ALASKA, ALBERTA, BRITISH COLUMBIA, COLORADO,

NEWFOUNDLAND. Other material studied: 21 M, 8 F (AEI, CNC, USNM): ALASKA, BRITISH COLUMBIA, NORTHWEST TERRITORIES.

ETYMOLOGY. *Lat-* (wide) + *-issimus* (-est), referring to the very wide postpetiole and 2nd abdominal tergum.

Endasys leptotexanus Luhman, sp. n.
(Fig. 6:A-D; Map 12)

MALE DIAGNOSIS. Medium large, 5.5-7.5 mm long; legs very slender (Fig. 6:C), Rs very long, clypeus white, mostly orange legs and abdomen, sometimes white on coxa I and trochanters I and II, punctation generally very dense with finely distinct pits, surface often appearing slightly rugulose, and very small 3rd tyloid on 12th flagellar segment; Alaska.

MALE DESCRIPTION. *White*: clypeus, mandible, sometimes coxa I apically, sometimes front and middle 1st trochanter apically, front and middle trochantelli entirely. *Yellow*: Scape, and sometimes coxae I and II yellow. *Orange*: Legs I and II more yellow-orange except sometimes white on coxae and trochanters, leg III except tibia apically and most of tarsomeres, abdomen except apically and sometimes basally. *Black*: Flagellum, tibia III apically, hind tarsomeres except basal segment, abdomen apically and sometimes basally. *Punctation* (Fig. 86:B,D-F): Face very finely rugulose or granular with very dense pits; frons with slightly rugulose, very dense, fine pits; temple with dense, indistinct pits; propleurum with very dense, distinct pits; mesopleurum impunctate centrally and slightly rugulose, otherwise rugulose with very dense, very fine, distinct pits; postpetiole and 2nd tergum distinctly mat. *Shape*: Clypeal margin not upturned (Fig. 78:A); flagellum moderately slender (Fig. 80:B), its segments longer than wide to apex; areola widely hexagonal, area dentipara narrowed and often elongate apically, apophysis very strong and toothlike, propodeum slightly sloped, carinae strong; 1st abdominal tergum moderately stout (Fig. 84:E), postpetiole squarish with strong dorsomedian carina 2nd tergum regular to widened; femur III slender (Fig. 79:A), ratio 0.18; radial cell very long, longer than 1.5 length of discal cell. *Other*: Wings tinted, 3rd tyloid on 12th flagellar segment very small, glumes sparse.

FEMALE DIAGNOSIS. Medium large, 7.5 mm long; orange scape, clypeus, legs, and abdomen; radial cell distinctly elongate; propodeal carinae moderately strong; femur III moderately slender (Fig. 79:B), ratio 0.22; clypeus without upturned margin (Fig. 78:A); punctation generally very dense, surfaces slightly rugulose; flagellum bicolored, distinctly narrowed on apical half.

FEMALE DESCRIPTION. *Orange*: Flagellum basally half, scape, clypeus, mandible, tegula, legs, and abdomen. *Black*: Flagellum on apical half. *Punctation* (Fig. 86:C-F): Face and frons finely rugulose with very dense pits; temple with dense, indistinct pits; propleurum with dense, fine, more or less distinct pits; mesopleurum with sparse, distinct pits and some rugulosity. *Shape*: Clypeal margin not upturned (Fig. 78:A); flagellum moderately stout and distinctly narrowed apically, basal 2 segments longer than wide (Fig. 80:E,H); areola widely hexagonal, area dentipara narrowed and often elongate apically, apophysis very strong and sharp, propodeal carinae moderately strong; 1st abdominal tergum moderately slender (Fig. 84:C), postpetiole gradually expanded, about 1.7 times as wide as long, sides very slightly converging apically, dorsomedian carina strong; femur III moderately slender (Fig. 79:B), ratio 0.22; radial cell distinctly elongate. *Other*: 3rd valvula moderately short, about as long as basal 3 hind tarsomeres; wings tinted

REMARKS. See discussion under *rugitexanus* and *rhyssotexanus*.

RANGE. Alaska. Collections in July.

MATERIAL EXAMINED. **Holotype**: male, Tsaina R., ALASKA, 16-viii-1973, H. and M. Townes (AEI). **Paratypes**: 3 M, 1 F (AEI): ALASKA. Other material studied: 1 M (AEI): ALASKA.

ETYMOLOGY. (Greek) *Lepto-* (slender) + *texanus* (the species), referring to its similarity to *texanus* except for the slender legs.

Endasys maculatus (Provancher)
(Fig. 41:A-D; Map 6)

Phygadeuon maculatus Provancher, 1875: 178, 182. **Holotype** cited by Gahan and
    Rohwer, 1918: 136; Townes, 1939: 94; and Barron, 1975: 392. References:
    Provancher, 1879: 67, 1882: 333, 353; 1883: 314, 775; 1886: 45. Dalla Torre, 1902:
    688.
Stylocryptus maculatus: Cushman, 1925: 389; 1928: 928.
Endasys (Endasys) maculatus: Townes, 1944: 212; Townes and Townes, 1951: 246.
Endasys maculatus: Carlson, 1979: 417.

MALE DIAGNOSIS. Elliptical, white patches on outer faces of tibiae; 3rd tyloid on flagellar segment 12; head coarsely and densely punctate; propleurum and mesopleurum rugulose and coarsely punctate. Only *mucronatus* similar but with white on scape, coxae, and trochanters; eastern half of United States and southern Canada.

MALE DESCRIPTION. *White*: Elliptical patches on outer faces of tibiae. *Black*: Scape, clypeus, mandible, tegula, abdomen, and legs. *Punctation* (Fig. 86:B,F): Face finely rugulose; frons with very dense, coarse pits; temple slightly rugulose with dense pits; propleurum and mesopleurum rugulose with coarse pits, central area of mesopleurum impunctate and rugulose; petiole wrinkled but shiny, 2d tergum wrinkled and shiny with variably sparse, fine pits. *Shape*: Clypeal margin not upturned (Fig. 78:A); flagellar segments longer than wide to apex; areola widely, irregularly hexagonal, or nearly arched; area dentipara very narrow, nearly triangular; petiole stout, postpetiole square; 2d tergum wide. *Other*: Propodeal carinae and apophysis strongly developed; short 3rd tyloid on flagellar segment 12; wings hyaline to slightly tinted.

FEMALE DIAGNOSIS. Large, 7-9 mm long; entirely black except white elliptical patches on tibiae dorsally; white on flagellar segments 3 or 4 to 9.

FEMALE DESCRIPTION. *White*: Flagellar segments 3 or 4-9, and elliptical patches on tibiae dorsally. *Black*: Flagellum basally and apically, scape, mandible, tegula, legs except tibiae dorsally, and abdomen. *Punctation* (Fig. 86:F): Face, frons, temple, and propleurum rugulose with very dense, coarse pits; mesopleurum rugulose with variably dense or sparse, coarse pits. *Shape*: Clypeal margin not upturned (Fig. 78:A); flagellum moderately stout, narrowed apically, basal 3 segments elongate (Fig. 80:E,I); areola arched, sometimes nearly trapezoidal, area dentipara distinctly narrowed, nearly triangular, apophysis distinct and sharp, propodeal carinae distinct; 1st tergum moderately stout, postpetiole wide, about twice as wide as long, dorsomedian carina distinct; femur III swollen, ratio 0.23-0.24; radial cell long, about 1.5 times longer than 2nd-discoidal cell, Rs straight. *Other*: 3rd valvula moderately short, about as long as basal 3 hind tarsomeres; wings slightly tinted.

REMARKS. This species differs from all others in the Nearctic Texanus Group by the white, elliptical patches on the tibiae dorsally. It is closely related to Palearctic *parviventris* (Gravenhorst) which also has these patches, but differs by its orange abdomen and mostly orange femur III. The femur III and abdomen of *maculatus* are black.

RANGE. Eastern North America from Minnesota across southern Canada south to Louisiana and Florida. Collections June through August.

MATERIAL EXAMINED. **Holotype**: female , QUEBEC, 1875, Provancher (UL). No paratypes. Other material studied: 21 M, 12 F (AEI, CNC, Dasch, FLDA, MCZ, UM, USNM): CONNECTICUT, MAINE, MARYLAND, MASSACHUSETTS, MICHIGAN, MINNESOTA, NEW HAMPSHIRE, NEW JERSEY, NEW YORK, NORTH CAROLINA, NOVA SCOTIA, OHIO, ONTARIO, SOUTH CAROLINA, VERMONT, WASHINGTON D.C.

ETYMOLOGY. *Maculatus* (spotted), referring to the white patches on the tibiae.

Endasys mucronatus (Provancher)
(Fig. 29:A-D; Map 20)

Phygadeuon mucronatus Provancher, 1879: 73; 1882: 334, 335, 353; 1883: 319, 775;
  1886: 46, 48, 49; Ashmead, 1900: 568. **Lectotype** designated by Barron, 1975: 511.
Melophron (!) *abdominalis* Ashmead, manuscript name, *in* Slosson, 1906: 324;
  synonymized by Cushman, 1922: 18; as nomen nudum by Carlson, 1979: 417.
Stylocryptus mucronatus: Cushman, 1922: 28; (1926) 1928: 928; Gobeil, 1937: 87.
Endasys (Endasys) mucronatus: Townes, 1944: 212; Townes and Townes, 1951: 246.
Endasys mucronatus: Carlson, 1979: 417.

MALE DIAGNOSIS. Large, 7-9 mm long; 3rd tyloid on flagellar segment 12; mostly black except white scape, coxae apically, trochanters, tibiae I and II dorsally, sometimes tibia III with white stripe dorsally; eastern half of United States and southern Canada.

MALE DESCRIPTION. *White*: Scape, tegula variably, coxae usually apically, trochanters usually, tibiae I and II dorsally, and often tibia III dorsally; sometimes white only on trochanters and tibiae I and II dorsally. *Orange*: Orange forms with abdomen except apically. *Black*: Flagellum, clypeus, mandible, coxae except apically, leg III except often tibia dorsally, usually abdomen. *Punctation* (Fig. 86:F): Face and frons finely rugulose with very dense pits; temple slightly rugulose with dense, more or less distinct pits;

propleurum rugulose with dense to sparse, coarse pits; mesopleurum impunctate centrally and rugulose, otherwise rugulose with very dense pits; 1st and 2nd terga smooth and shiny. *Shape*: Clypeal margin not or only very slightly upturned (Fig. 78:B); flagellum moderately slender (Fig. 80:B), its segments longer than wide to apex; areola nearly rectangular, widely hexagonal, area dentipara distinctly narrowed, apophysis very sharp and strong, propodeal carinae strong, surfaces coarse between carinae; 1st tergum moderately slender (Fig. 84:F), postpetiole wide or square, 2nd tergum wide to regular; femur III moderately swollen (Fig. 79:C), ratio about 0.23. *Other*: 3rd tyloid on flagellar segment 12, sometimes faint 4th tyloid on segment 13; wings hyaline to slightly tinted; 5th abdominal sternum usually complete; glumes dense.

REMARKS. The more common color of the male abdomen is black. Males with orange abdomens are less than 5% of collected specimens; however, half of a reared series in the CNC are orange males. Females always have an orange abdomen.

FEMALE DIAGNOSIS. Large, 9-10 mm long; white flagellar segments 4 or 5 to 10; punctation of head and thoracic pleura generally rugulose with very dense, coarse pits; apophysis strong and toothlike; abdomen orange, legs mostly black except front and middle tibiae distinctly white dorsally, and sometimes also hind tibia dorsally; 3rd valvula long, about as long as hind tarsomeres.

FEMALE DESCRIPTION. *White*: Flagellar segments 4 or 5 to 10 or 11, often 1st trochanters apically and trochantelli basally, often coxae apically, tibiae I and II dorsally, sometimes tibia III dorsally. *Orange*: Abdomen. *Black*: Flagellum except annulus, clypeus, mandible, legs except white on coxae, trochanters, and tibiae dorsally; sometimes 1st tergum black basally, and sometimes legs entirely black except tibiae I and II dorsally. *Punctation* (Fig. 86:F): Head and thoracic pleura rugulose with very dense, coarse pits. *Shape*: Clypeal margin not upturned (Fig. 78:A); flagellum moderately slender narrowed apically, basal 3 segments elongate (Fig. 80:E,I); areola broadly hexagonal, area dentipara narrowed, apophysis strong and toothlike, vertically projecting, propodeal carinae strong; 1st tergum moderately slender (Fig. 84:F), postpetiole moderately widened, about twice as wide as long; femur III moderately swollen (Fig. 79:D), ratio 0.24; Rs straight, very slightly bowed (Fig. 82:A). *Other*: 3rd valvula long, about as long as hind tarsomeres; wings tinted.

REMARKS. See discussion under *texanus* and *taiganus*.

RANGE. Eastern half of the United States and southern Canada from Minnesota to Quebec, southward to Louisiana and Florida. Collections May through October, most commonly collected late summer and early fall.

HOST. Argidae: *Arge pectoralis* (Leach), birch sawfly (CNC); *A. clavicornis* (F.) (Gobeil, 1937: 87).

MATERIAL EXAMINED. **Lectotype**: female, Ste. Hyacinthe, QUEBEC, 1879, Provancher (UL). Paralectotypes: none so labelled by Barron (1975: 415). Other material studied: 145 M, 47 F (AEI, CAS, CNC, Dasch, FLDA, USNM): ALBERTA, CONNECTICUT, FLORIDA, GEORGIA, LOUISIANA, MANITOBA, MARYLAND, MASSACHUSETTS, MICHIGAN, MINNESOTA, NEW BRUNSWICK, NEW HAMPSHIRE, NEW JERSEY, NEW YORK, NORTH CAROLINA, NOVA SCOTIA, OHIO, ONTARIO, PENNSYLVANIA,

QUEBEC, RHODE ISLAND, SOUTH CAROLINA, VIRGINIA, WASHINGTON D.C.,WEST VIRGINIA.

ETYMOLOGY. *Mucron-* (sharp point) + *-atus* (-ed), referring to the strong, toothlike apophysis.

Endasys oregonianus Luhman, sp. n.
(Fig. 28:A-D; Map 15)

MALE DIAGNOSIS. Large, 7.5-9.5 mm long; 3rd tyloid on flagellar segment 12 small to faint; propodeal carinae strong, apophysis strong and toothlike; frons and face finely rugulose with very dense pits; orange scape, legs, and abdomen; black clypeus, hind tarsomeres and apex of abdomen; western United States and Canada.

MALE DESCRIPTION. *Orange*: Scape, tegula, legs except hind tarsomeres, abdomen except apex. *Black*: Flagellum, clypeus, mandible, hind tarsomeres, and apex of abdomen; rarely on scape, tegula, coxae, and trochanters. *Punctation* (Fig. 86:C-F): Face and frons finely rugulose with very dense pits; temple with dense, indistinct pits; propleurum slightly rugulose with very dense to sparse, more or less distinct pits; mesopleurum impunctate and slightly rugulose centrally, otherwise rugulose and densely pitted; 1st and 2nd terga very slightly mat and sometimes very slightly wrinkled. *Shape*: Clypeal margin not upturned (Fig. 78:A); flagellum moderately slender (Fig. 80:B), its segments longer than wide to apex; areola widely hexagonal, area dentipara narrowed, apophysis strong and toothlike, propodeal carinae strong; 1st tergum slender, postpetiole longer than wide, dorsomedian carina strong, 2nd tergum regular; femur III slender (Fig. 79:A), ratio 0.20; Rs often very slightly bowed (Fig. 82:A), otherwise straight. *Other*: 3rd tyloid on flagellar segment 12 short or faint, tyloids on segments 10-11 moderately long; wings more or less hyaline or tinted; 5th abdominal sternum complete medially.

FEMALE DIAGNOSIS. Large, 7-8.5 mm long; orange scape, tegula, legs, and abdomen, hind tarsomeres blackish orange, clypeus black; 3rd valvula moderately long, as long as basal 4 hind tarsomeres; apophysis very strong and toothlike; femur III slender (Fig. 79:B), ratio 0.22-0.23; flagellum blackish orange, slender, narrowed apically (Fig. 80:E); punctation of head and thoracic pleura generally finely rugulose with very dense pits.

FEMALE DESCRIPTION. *Orange*: Flagellum blackish orange, scape, mandible, tegula, legs except hind tarsomeres more blackish orange, abdomen. *Black*: Sometimes flagellum and scape, clypeus, hind tarsomeres blackish orange. *Punctation* (Fig. 86:D-F): Face finely rugulose with very dense pits; frons rugulose with very dense, coarse pits; temple with dense, more or less distinct pits; propleurum with variably sparse to dense, distinct pits, surface a little rugulose; mesopleurum strongly rugulose with dense pits. *Shape*: Clypeal margin not upturned (Fig. 78:A); flagellum slender, narrowed apically, basal 3 segments elongate (Fig. 80:E,I); areola nearly hexagonal and slightly arched, area dentipara narrowed, apophysis strong and toothlike, propodeal carinae strong; 1st tergum moderately slender (Fig. 84:C), postpetiole modertely widened, about 1.5 times as wide as long, dorsomedian carina distinct; femur III slender (Fig. 79:B), ratio 0.22-0.23; Rs straight, often very slightly bowed (Fig. 82:A). *Other*: 3rd valvula moderately long, as long as basal 4 hind tarsomeres; wings tinted.

REMARKS. This species is closely related to *rubescens*. Male *oregonianus* is distinguished by the black clypeus, orange hind tibia without black apically, black hind tarsomeres, and head and thorax without orange patches. Females are distinguished by the blackish orange flagellum without a white annulus on segments 5-9, and never have orange patches on the head and thorax.

RANGE. Pacific Northwest from British Columbia to northern California, westward to Idaho and Colorado. Collections May, July, and August.

MATERIAL EXAMINED. **Holotype**: male, Mt. Rainier, 2900 ft., OREGON, 28-vii-1940, H. and M. Townes (AEI). **Paratypes**: 58 M, 8 F (AEI, CAS, CNC, USNM): BRITISH COLUMBIA, CALIFORNIA, IDAHO, OREGON, WASHINGTON. Other material studied: 2 M (AEI, CAS): CALIFORNIA.

ETYMOLOGY. From Oregonian Biotic Province, referring to the range of the type series.

Endasys pseudocallistus Luhman, sp. n.
(Fig. 64:A-D; Map 8)

MALE DIAGNOSIS. Medium large, 6-8 mm long; face very finely granular with very dense, very fine pits; frons finely rugulose with very dense, fine pits; 1st abdominal tergum moderately stout (Fig. 84:E), postpetiole squarish, mat and slightly wrinkled; propodeum distinctly short with coarsely distinct carinae, apophysis weak; 2nd tergum wide and mat; clypeus orangish black, scape yellow, leg I yellowish, legs II and III orange except apices of femur III and tibia III black, abdomen bicolored, orange on terga 1-3 and basally 4-5, black on terga 4 and 5 apically to apex of abdomen; flagellum moderately stout (Fig. 80:C), segments squarish on apical half; eastern half of northern United States and southern Canada.

MALE DESCRIPTION. *Yellow*: Flagellum ventrally, scape, leg I, trochanters except often more orangish basally. *Orange*: Clypeus orangish black, mandible, tegula, coxae II and III, trochanters basally, femur III except dorso-apically, tibia III except apically, hind tarsomeres, abdominal terga 1-3, basally on terga 4 and 5. *Black*: Clypeus orangish black, femur III dorso-apically, tibia III apically, hind tarsomeres, apically on terga 4 and 5 to apex of abdomen; sometimes 1st tergum blackish basally. *Punctation* (Fig. 86:B-C,E-F): Face finely granular with very dense, fine pits; frons with very dense, fine pits, surface appearing a little granular; temple with dense, indistinct pits; propleurum with variably dense, variably sized pits, surface smooth to rugulose; mesopleurum impunctate centrally and a little rugulose, otherwise densely punctate and rugulose; 1st and 2nd terga mat, 1st tergum slightly wrinkled. *Shape*: Clypeal margin sharp and slightly upturned (Fig. 78:B); flagellum moderately stout (Fig. 80:C), its segments squarish on apical half; areola widely hexagonal, nearly quadrangular, area dentipara narrowed, apophysis weak, propodeum distinctly shortened with coarse, distinct carinae; 1st tergum moderately stout (Fig. 84:E), postpetiole squarish, 2nd tergum wide; femur III moderately slender (Fig. 79:B), ratio 0.20-0.22; Rs straight. *Other*: Tyloids (2) short but prominent (Fig. 87:D); wings hyaline; 5th abdominal sternum more or less complete medially.

FEMALE DIAGNOSIS. Medium large, 6.5 mm long; face and clypeus very finely granular with very dense pits; frons slightly rugulose to granular with very dense, fine pits; propodeum distinctly shortened, carinae coarsely distinct, apophysis distinct and wide; 1st tergum stout (Fig. 84:A), postpetiole wide; flagellum bicolored, yellowish basally, moderately slender, narrowed apically, basal 3 segments distinctly elongate (Fig. 80:E,I); I and II yellow, leg III mostly orange, abdomen orange except most of 1st tergum.

FEMALE DESCRIPTION (based on one specimen). *Yellow*: Flagellum basal half, scape, mandible, tegula, legs I and II, hind trochanters. *Orange*: Leg III except trochanters ventrally, abdomen except most of 1st tergum. *Black*: Flagellum basal half, clypeus, 1st tergum except apically. *Punctation* (Fig. 86:B,D,F): Face very finely granular with very dense pits; frons a little rugulose or granular with very dense, fine pits; temple with dense, more or less distinct pits; propleurum rugulose with very dense pits; mesopleurum with variably dense, more or less distinct pits, surface mostly rugulose. *Shape*: Clypeal margin not or only very slightly upturned (Fig. 78:B); flagellum moderately slender, narrowed apically, basal 3 segments distinctly elongate (Fig. 80:E,I); areola widely hexagonal, area dentipara distinctly narrowed, short, apophysis distinct and wide, propodeum short; 1st tergum stout (Fig. 84:A), postpetiole wide, 1.75 times as wide as long; femur III moderately swollen (Fig. 79:D), ratio 0.28; radial cell long, about 1.5 times as long as 2nd-discoidal cell, Rs mostly straight. *Other*: 3rd valvula long, about as long as hind tarsomeres; wings a little tinted.

REMARKS. See discussion under *latissimus*.

RANGE. Minnesota, Michigan, Ontario, Pennsylvania, New York, and Maine. Collections May through July.

MATERIAL EXAMINED. **Holotype**: male, Ann Arbor, MICHIGAN, 29-v-1959, H. and M. Townes (AEI). **Paratypes**: 9 M, 1 F (AEI, Dasch, UM): MAINE, MICHIGAN, MINNESOTA, NEW YORK, PENNSYLVANIA.

ETYMOLOGY. (Greek) *Pseudo-* (false) + *callistus*, referring to resemblance to *callistus*.

Endasys rhyssotexanus Luhman, sp. n.
(Fig. 47:A-D; Map 8)

MALE DIAGNOSIS. Medium large, 5.5-6.5 mm long; propodeum short with strongly wrinkled carinae, apophysis very strong; 3rd short tyloid on flagellar segment 12; 1st tergum moderately slender (Fig. 84:F), postpetiole a little longer than wide; pale orange clypeus, scape, legs, and abdomen, with black apically on femur III, tibia III, and abdomen, dark forms also with black coxae and terga 3-6; femur III moderately slender (Fig. 79:B), ratio 0.21-0.22; Alaska, British Columbia, and Colorado.

MALE DESCRIPTION. *Orange*: Scape, clypeus, mandible, tegula variably, coxae variably, trochanters, femora except femora II and II apically, tibiae except tibia III apically, most of abdomen except apex and variably on terga 1 and 3-6. *Black*: Tegula variably, coxae variably, femora II and III apically, tibia III apically, most of hind tarsomeres, abdomen apically and variably on terga 3-6. *Punctation* (Fig. 86:A-D): Face finely rugulose with very dense pits; frons with dense, very finely distinct pits; temple with dense hairs, pits indistinct; propleurum very densely punctate to impunctate, pits more or less distinct, surface slightly rugulose; mesopleurum impunctate centrally and rugulose, otherwise very densely punctate and rugulose; 1st and 2nd terga mat, 3rd tergum slightly mat on basal half. *Shape*: Clypeal margin weakly upturned (Fig. 78:B); flagellum moderately stout (Fig. 80:C), its segments squarish on apical half; areola widely hexagonal, nearly quadrangular, area dentipara distinctly narrowed, apophysis strong, projecting apically, propodeum rugulose with strong carinae; 1st tergum moderately slender (Fig. 84:F), postpetiole squarish to a little longer than wide, 2nd tergum regular; femur III moderately slender (Fig. 79:B), ratio about 0.21-0.22; radial cell mostly straight except apically (Fig. 82:C). *Other*: Shorter, 3rd tyloid on flagellar segment 12; wings more or less hyaline; 5th abdominal sternum complete; glumes sparse (Fig. 88:C).

FEMALE DIAGNOSIS. Medium large, 6.5-7 mm long; propodeum short with strong, coarse carinae, area dentipara distinctly narrowed, apophysis strong, narrow and toothlike, projecting somewhat posteriorly; flagellum moderately stout (Fig. 80:C), narrowed apically, mostly black, but pale orangish baso-ventrally, and sometimes dorsally on segments 5-9; 3rd valvula moderately long, about as long as basal 3 hind tarsomeres; 1st tergum moderately stout (Fig. 84:B), postpetiole moderately widened, about 1.5 times as wide as long; hind femur III moderately slender (Fig. 79:B), ratio about 0.24; mostly pale orange scape, clypeus, legs, and abdomen except apices of femur III, tibia III, and abdomen.

FEMALE DESCRIPTION. *Orange*: Flagellum baso-ventrally, sometimes flagellar segments 5-9 more orangish, scape, clypeus, mandible, tegula, coxae except coxae II and III dorsally, trochanters, femora I and II, tibiae I and II except tibia II dorsally, femur III except dorsally and apically, tibia III except apically, abdomen except apically. *Black*: Flagellum dorsally and ventrally on apical half except sometimes segments 5-9 orangish, femora II and III dorsally, femur III and tibia III apically, hind tarsomeres, abdomen apically orangish black. *Punctation* (Fig. 86:A-D): Face finely rugulose with very dense pits; frons with dense, finely distinct pits, surface smooth; temple indistinctly punctate, hairs dense; propleurum with mostly dense, very fine, more or less distinct pits; mesopleurum with variably dense to sparse punctation, more or less distinct pits, some rugulosity. *Shape*: Clypeal margin not upturned (Fig. 78:A); flagellum moderately slender, apically narrowed, basal 3 segments elongate (Fig. 80:E,I); areola wide hexagonal, area dentipara distinctly narrowed, apophysis strong, projecting somewhat posteriorly, propodeal carinae distinct; 1st tergum moderately stout (Fig. 84:B), postpetiole moderately widened, width about 1.5 times length; femur III moderately slender (Fig. 79:B), ratio 0.23-0.25; Rs straight. *Other*: 3rd valvula moderately long, as long as basal 3 hind tarsomeres; wings a little tinted.

REMARKS. This species is probably most closely related to *leptotexanus*. Males are distinguished by the orange scape, clypeus, coxae, and trochanters, and black on the hind leg and abdominal terga 2-6. Females differ by the orange flagellum, not bicolored, and black apically on the hind femur and tibia.

RANGE. Alaska and British Columbia to Colorado. Collections June through August.

MATERIAL EXAMINED. **Holotype**: male, Stone Mt. Pk., 3800+ ft., BRITISH COLUMBIA, 13-vii-1973, H. and M. Townes (AEI). **Paratypes**: 7 M, 3 F (AEI, Dasch): ALASKA, BRITISH COLUMBIA, CALIFORNIA, COLORADO, OREGON. Other material studied: 1 M (AEI): BRITISH COLUMBIA.

ETYMOLOGY. *Rhysso-* (wrinkled) + *texanus*, referring to the similarity to *texanus*, but with wrinkled abdominal terga 1-3 and propodeal carinae in the male.

Endasys rubescens Luhman, sp. n.
(Fig. 7:A-D; Map 19)

MALE DIAGNOSIS. Medium large, 6-7.5 mm long; clypeal margin sharp and more or less upturned (Fig. 78:B), face finely rugulose with very dense pits, frons very densely and finely punctate, often slightly rugulose; often orange patches on head or thorax, legs and abdomen mostly dark orange; 3 tyloids on flagellar segments 10-12, flagellum moderately slender, segments longer than wide to apex (Fig. 80:B); propodeum rugulose with strong carinae, area dentipara narrowed with strong apophysis; postpetiole and 2nd tergum distinctly mat and slightly wrinkled, 1st tergum with very strong dorsomedian carina; scape, clypeus, and coxae variably orange or black; wings blackish, radial cell long, Rs straight; femur III slender (Fig.79:A), ratio about 0.18; West Coast.

MALE DESCRIPTION. *Orange*: Usually scape and clypeus, mandible, tegula, legs except tibia III apically and hind tarsomeres, abdomen except usually apically, often orange patches on head or thorax. *Black*: Flagellum, sometimes scape and clypeus, usually hind tibia apically, hind tarsomeres; dark forms with black coxae and 1st-trochanters. *Punctation* (Fig. 86:B-E): Face very finely rugulose with very dense pits; frons with very dense, fine pits, often slightly rugulose; temple with more or less dense, indistinct pits; propleurum with dense to sparse punctation, more or less distinct pits; mesopleurum mostly impunctate centrally or sparsely punctate with distinct pits, otherwise variably punctate with distinct pits, and slightly rugulose; postpetiole and 2nd tergum distinctly mat and slightly wrinkled. *Shape*: Clypeus with sharp edge more or less upturned (Fig. 78:C); flagellum moderately slender (Fig. 80:B), segments longer than wide to apex; areola broadly to widely hexagonal, area dentipara narrowed, apophysis strong, propodeum rugulose with strong carinae; 1st abdominal tergum moderately stout (Fig. 84:E), postpetiole longer than wide, dorsomedian carina very strong, 2nd tergum more or less regular; femur III slender (Fig. 79:A), ratio about 0.18; radial cell long, Rs straight. *Other*: 3 unequal-length tyloids on flagellar segments 10-12, wings blackish, 5th abdominal sternum complete, and cranial suture more or less distinct below median ocellus.

FEMALE DIAGNOSIS. Medium large, 7-7.5 mm long; clypeal margin slightly upturned medially (Fig. 78:B); flagellum bicolored, moderately slender, narrowed apically (Fig. 80:E); head, thorax, and legs mostly orange; areola nearly trapezoidal, area dentipara distinctly narrowed, apophysis strong; 1st abdominal tergum stout (Fig. 84:A), postpetiole about twice as wide as long; 3rd valvula short or moderately short, about as long as basal 3 hind tarsomeres; wings blackish.

FEMALE DESCRIPTION. *Yellowish orange*: Flagellar segments 5-9, legs I and II. *Orange*: Nearly entire head, thorax, abdomen, legs, scape and basal 4 flagellar segments. *Black*: Flagellum on apical half, marginal areas of thorax. *Punctation* (Fig. 86:B,D-E): Face, frons, and propleurum rugulose with very dense, very finely distinct pits; temple with dense, fine, more or less distinct pits; mesopleurum rugulose, punctation variably dense but distinct. *Shape*: Clypeal margin slightly upturned medially (Fig. 78:B); flagellum moderately slender, narrowed apically (Fig. 80:E), first 3 basal segments a little elongate or square; areola irregularly hexagonal, nearly trapezoidal, area dentipara distinctly narrowed and slightly elongate, apophysis strong, propodeum finely rugulose with distinct carinae; 1st tergum stout (Fig. 84:A), postpetiole greater than twice as long as wide, dorsomedian carina very strong; femur III moderately swollen (Fig. 79:D), ratio about 0.24. *Other*: 3rd valvula moderately short, about as long as basal 3 hind tarsomeres; wings blackish.

REMARKS. See discussion under *oregonianus*.

RANGE. West Coast from Baja California to Oregon and east to Idaho. Collections March through August: March through May in California; June through August in Oregon and Idaho.

MATERIAL EXAMINED. **Holotype**: male, Pinehurst, OREGON, 19-vi-1978, H. and M. Townes (AEI). **Paratypes**: 28 M, 3 F (AEI, CAS, Dasch, UCR): BAJA CALIFORNIA NORTE, CALIFORNIA, IDAHO, OREGON. Other material studied: 2 M, 2F (AEI, OSU): CALIFORNIA, WASHINGTON.

ETYMOLOGY. *Rub-* (red) + *-escens* (becoming), referring to the orange on the head and thorax.

Endasys rugitexanus Luhman, sp. n.
(Fig. 23:A-D; Map 13)

MALE DIAGNOSIS. Medium large, 6-8 mm long; 3 tyloids on flagellar segments 10-12; white scape, clypeus, coxae apically, and trochanters; 1st abdominal tergum moderately stout (Fig. 84:E), mat and slightly wrinkled, postpetiole square to longer than wide, dorsomedian carina strong, 2nd tergum mat and slightly wrinkled; face and frons finely rugulose with very dense pits; coxae basally orange or black, femur III orange with black apex or entirely black, tibia III orange with black apex, and abdomen weakly bicolored, some blackish on terga 6-apex, sometimes terga 4-apex; mostly eastern half of northern United States and southern Canada.

MALE DESCRIPTION. *White*: Scape, clypeus, mandible, coxae apically, trochanters except hind trochanters sometimes more yellowish, femora I and II dorsally. *Yellow*: Sometimes flagellum more yellowish ventrally, tegula, femora I and II, tibiae I and II except dorsally, hind trochanters sometimes more yellowish. *Orange*: Usually coxae I and II basally, coxa III, femur III except apical third, tibia III except apically, often hind basitarsus, usually abdominal terga 6-apex. *Black*: Usually flagellum at least dorsally, femur III at least apical third, tibia III apically, hind basitarsus, and usually terga 6-apex; sometimes coxae mostly black basally, femur III mostly black, abdomen basally and on terga 4-apex. *Punctation* (Fig. 86:B,D-F): Face finely rugulose with very dense pits; frons with dense, distinct pits; temple with dense, indistinct pits; propleurum with, dense to sparse, variably distinct pits, often coarse and rugulose; mesopleurum impunctate centrally and usually rugulose, otherwise mostly rugulose with dense, coarse pits; 1st and 2nd terga mat and slightly wrinkled. *Shape*: Clypeus without margin upturned (Fig. 78:A); flagellum moderately slender; areola widely hexagonal or nearly rectangular, area dentipara distinctly narrowed and elongate apically, apophysis distinct and sharp, propodeal carinae strong, mostly rugulose between carinae; 1st tergum usually stout to moderately stout (Fig. 84:E), postpetiole square to longer than wide, dorsomedian carina strong; femur III moderately slender (Fig. 79:B), ratio about 0.21; radial cell long, Rs straight, areolet distinctly narrowed. *Other*: 3rd tyloid on flagellar segments 12 shorter than tyloids on segments 10-11; wings nearly hyaline; 5th abdominal sternum complete medially (Fig. 81:A).

FEMALE DIAGNOSIS. Medium long, 7 mm long; distinct white annulus on flagellar segments 4 to 9 or 10; face, frons, propleurum, and mesopleurum finely rugulose with very dense pits; areola nearly quadrangular, a little arched, area dentipara distinctly narrowed, apophysis distinct, propodeum short; 1st abdominal tergum stout, postpetiole wide, sides not distinctly parallel; abdomen weakly bicolored, black apically, otherwise orange; legs orange except femur III and tibia III apically, and hind tarsomeres entirely.

FEMALE DESCRIPTION. *White*: Flagellar segments 4-9 or 10. *Orange*: Scape, clypeus often, tegula, legs except femur III and tibia III apically and hind tarsomeres, abdomen except apically. *Black*: Flagellum except annulus, usually clypeus and mandible, femur III and tibia III apically, hind tarsomeres, abdomen apically. *Punctation* (Fig. 86:F): Face, frons, propleurum, mesopleurum, finely rugulose with very dense pits; temple slightly rugulose with dense pits. *Shape*: Clypeus without margin upturned (Fig. 78:A); flagellum moderately slender, narrowed apically (Fig. 80:E), basal 3 segments elongate; areola widely hexagonal, nearly quadrangular, a little arched, area dentipara distinctly narrowed, apophysis distinct, propodeal carinae distinct; 1st tergum stout (Fig. 84:A), postpetiole wide, but sides slightly curved; femur III moderately swollen (Fig. 79:D), ratio about 0.26; radial cell long, Rs straight. *Other*: 3rd valvula long, as long as basal 4 hind tarsomeres; wings tinted.

REMARKS. This species is related to *leptotexanus, angularis,* and *albitexanus.* Males of *rugitexanus* are distinguished from *leptotexanus* by the moderately slender femur III (ratio greater than 0.21), femur III orange with at least apex black, and abdominal terga 4-6 partly black; from *angularis* by having 3 tyloids and propodeal carinae strong, but not coarse and wrinkled; from *albitexanus* by having black hind tarsomeres, not whitish, and the 2nd abdominal tergum slightly mat and wrinkled, not smooth and shiny. Female *rugitexanus* is distinguished from both *albitexanus* and *angularis* by the femur III more swollen (ratio 0.24-0.26), 3rd valvula longer (about as long as basal 4 hind tarsomeres), and the orange abdomen with black apically. Female *albitexanus* also differs by having whitish hind tarsomeres. Female *rugitexanus* differs from *leptotexanus* by the white flagellar segments 5-9 and black on the abdomen apically.

RANGE. Alaska to North Dakota and Minnesota, eastward to New Brunswick and New York and southward to West Virginia and Maryland. Collections May and June through September and October.

MATERIAL EXAMINED. **Holotype**: male, Plummers Is., MARYLAND, 4-x-1912, R.A. Cushman (USNM). **Paratypes**: 39 M, 17 F (AEI, AMNH, CNC, Dasch, INF, MCZ, UM, USNM, Yu): ALASKA, MAINE, MARYLAND, MICHIGAN, MINNESOTA, NEW BRUNSWICK, NEW YORK, NORTH DAKOTA, OHIO, ONTARIO, PENNSYLVANIA, WASHINGTON D.C., WEST VIRGINIA, WISCONSIN, VERMONT, VIRGINIA. Other material studied: 2 M, 3 F (AEI, USNM): MARYLAND, MICHIGAN.

ETYMOLOGY. *Rugi-* (rough) + *texanus* (species name), referring to similarity to *texanus* except for mat and wrinkled 1st and 2nd abdominal terga.

Endasys rugosus Luhman, sp. n.
(Fig. 53:A-D; Map 11)

MALE DIAGNOSIS. Medium large, 6-7 mm long; face rugulose with very dense, coarse pits; clypeus black, margin not upturned (Fig. 78:A); frons and propleurum a little rugulose with dense pits; legs yellow-orange except femur III and tibia III apically, and hind tarsomeres entirely; abdomen mostly orange, blackish on terga 4-apex; propodeum rugose between strong carinae, area dentipara narrowed, apophysis distinct; flagellum stout (Fig. 80:D) and mostly orange; northern United States and southern Canada.

MALE DESCRIPTION. *Orange*: Most of flagellum but darker dorsally, scape, clypeus sometimes, mandible, tegula, legs I and II more yellow-orange, leg III except femur dorso-apically and tibia apically, hind tarsomeres, abdomen except apically, sometimes on apically on terga 4-apex. *Black*: Clypeus usually, femur III usually dorso-apically, sometimes entirely, tibia III apically, hind tarsomeres more orangish black, abdomen apically and sometimes blackish apically on terga 4-apex; dark forms also with black flagellum, scape, tegula, coxae, and trochanters. *Punctation* (Fig. 86:A,D-F): Face rugulose with very dense, coarse pits; frons slightly rugulose with very dense pits; temple slightly rugulose with dense, more or less distinct pits; propleurum with dense to sparse, more or less coarse pits, surface a little rugulose at margins; mesopleurum impunctate centrally, otherwise with mostly dense, coarse pits, surface slightly rugulose; 1st and 2nd terga generally smooth and shiny. *Shape*: Clypeal margin not upturned (Fig. 78:A); flagellum stout or moderately stout, its segments squarish or slightly longer than wide on apical half; areola widely hexagonal, area dentipara distinctly narrowed, apophysis distinct, propodeum coarse between strong carinae; 1st tergum stout to moderately stout, postpetiole squarish or slightly wider than long, 2nd tergum wide; femur III moderately swollen (Fig. 79:C), ratio 0.22-0.24; radial cell long, Rs straight (Fig. 82:B). *Other*: Wings tinted, 5th abdominal sternum complete apically (Fig. 81:B), glumes moderately dense (Fig. 88:B), areolet a little widened.

FEMALE DIAGNOSIS. Medium large, 5-7 mm long; white on flagellar segments 4 or 5 to 9; face shiny and rugulose with dense, more or less coarse pits; clypeus black, convex, margin not upturned; abdomen orange with blackish on terga 4-apex; 3rd valvula moderately long, as long as basal 4 hind tarsomeres; coxae blackish orange, trochanters yellow-orange, femur III mostly black, tibia III yellowish orange with blackish apex; propodeum short and broad, area dentipara narrowed, apophysis distinct; 1st tergum stout (Fig. 84:A), nearly wedge-shaped, postpetiole widened to twice its length; flagellum moderately slender, very slightly narrowed apically, basal 3 segments short (Fig. 80:H).

FEMALE DESCRIPTION. *Orange*: Scape, coxae blackish orange, trochanters more yellow-orange, femora I and II blackish orange, femur III more basally, tibiae yellow-orange except tibia III apically, hind tarsomeres blackish orange, abdomen except apically on terga 4-apex. *Black*: Flagellum except annulus, sometimes basal 3 segments more brownish, clypeus, most of mandible, tegula, femur III except often basally, tibia III apically, often hind tarsomeres more blackish, abdomen blackish on terga 4-apex. *Punctation* (Fig. 86:D-F): Face, frons, and propleurum a little rugulose with dense, more or less coarse pits, surface shiny; temple slightly rugulose with moderately dense with more or less indistinct pits; mesopleurum with variably dense to sparse, distinct pits, surface with some rugulosity. *Shape*: Clypeus convex, margin not upturned (Fig. 78:A); flagellum moderately slender, very slightly narrowed apically, basal 3 segments short (Fig. 80:E,H); areola widely hexagonal, area dentipara narrowed, apophysis distinct, propodeum short and broad; 1st tergum distinctly stout (Fig. 84:A), nearly wedge-shaped, postpetiole gradually widened to twice the length; femur III moderately swollen (Fig. 79:D), ratio 0.26; radial cell, distinctly longer than 2nd-discoidal cell, Rs straight. *Other*: 3rd valvula moderately long, as long as basal 4 hind tarsomeres; wings distinctly tinted.

REMARKS. This species is placed in the Texanus Group with reservations. Males differ from other species in the group by the lack of white on the appendages, have 2 tyloids, and punctation is coarser and unevenly spaced. It is placed in this group, however, because of the strong, rugulose carinae of the propodeum, the strong apophysis, and the flagellum of the female with white on segments 5-9.

RANGE. Southern Canada and northern United States from British Columbia and Oregon eastward to New Brunswick, and southward to Maryland. Collections May through July.

MATERIAL EXAMINED. **Holotype**: male, Ann Arbor, MICHIGAN, 27-v-1959, H. and M. Townes (AEI). **Paratypes**: 58 M, 8 F (AEI, CNC, Dasch): BRITISH COLUMBIA, MANITOBA, MARYLAND, MICHIGAN, NEW BRUNSWICK, NEW JERSEY, NEW YORK, OREGON, PENNSYLVANIA. Other material studied: 26 M (AEI, Dasch, UM): COLORADO, MARYLAND, MINNESOTA, NEW BRUNSWICK, PRINCE EDWARD ISLAND, VERMONT, WEST VIRGINIA.

ETYMOLOGY. *Rugosus* (wrinkled), referring to rugulosity of the head, thoracic pleura, and especially the propodeum.

Endasys taiganus Luhman, sp. n.
(Fig. 39:A-D; Map 14)

MALE DIAGNOSIS. Large, 7-9 mm long; abdomen orange except basally and apically, black scape and clypeus, tibiae I and II white dorsally, coxae mostly black, whitish on apical margin, 1st-trochanters blackish, trochantelli whitish basally, femur III and tibia III black except latter orange basally; sometimes very small 3rd tyloid on 12th flagellar segment; area dentipara distinctly narrowed, apophysis strong and toothlike; Alaska and British Columbia, to South Dakota and Colorado.

MALE DESCRIPTION. *White*: Usually apical margin of 1st-trochanters, trochantelli basally, tibiae I and II dorsally, sometimes coxae on apical margin, rarely scape. *Yellow*: Femora I and II more or less apically, tibiae I and II more basally and ventrally. *Orange*: Tibia III basally, abdomen except basally and apically, often terga 2-6 laterally, sometimes femur III more orange basally. *Black*: Scape usually, clypeus, mandible, tegula, coxae except sometimes apical margin, 1st-trochanters except apical margin, trochantelli apically, femora I and II except apically more yellowish, tibiae I and II often apically blackish, femur III, tibia III except basally, hind tarsomeres, abdomen basally and apically, often blackish laterally on terga 2-6, sometimes terga 4-6 blackish apically. *Punctation* (Fig. 86:C,E-F): Face and frons finely rugulose with very dense pits; temple with dense, indistinct pits, propleurum with dense to sparse, distinct pits, surface little rugulose; mesopleurum impunctate and slightly rugulose centrally, otherwise rugulose with dense, distinct pits; 1st and 2nd terga weakly mat and weakly wrinkled. *Shape*: Clypeus without upturned margin (Fig. 78:A); flagellum moderately stout (Fig. 80:C), its segments squarish on apical half; areola widely hexagonal, nearly quadrangular, area dentipara distinctly narrowed, apophysis strong and toothlike, propodeal carinae strong; 1st tergum moderately stout (Fig. 84:E), postpetiole wider than long, 2nd tergum wide; femur III moderately slender (Fig. 79:B), ratio about 0.21; Rs straight, very slightly bowed (Fig. 82:A). *Other*: Usually 2 tyloids on flagellar segments 10-11, sometimes very small 3rd on 12th flagellar segment; wings tinted; 5th abdominal sternum complete medially; glumes moderately dense (Fig. 88:B).

FEMALE DIAGNOSIS. Large, 7-8 mm long; White flagellar segments 5-9, area dentipara distinctly narrowed, apophysis strong and toothlike, tibiae I and II white dorsally, black scape and clypeus, legs mostly black except tibia III orangish basally, abdomen entirely orange.

FEMALE DESCRIPTION. *White*: Flagellar segments 4 or 5 to 9, tibiae I and II dorsally. *Yellow*: Femora I and II except apically, tibiae I and II except dorsally and apically. *Orange*: Tibia III except apically, hind tarsomeres, and abdomen. *Black*: Flagellum except annulus, scape, clypeus, mandible, tegula, coxae, trochanters, femora except femora I and II apically, tibiae I and II except dorsally and apically, femur III, tibia III apically, hind tarsomeres. *Punctation* (Fig. 86:F): Face, frons, and propleurum rugulose with very dense pits; temple rugulose with dense pits; mesopleurum rugulose with dense pits. *Shape*: Clypeal margin not upturned (Fig. 78:A); flagellum moderately slender, narrowed apically, basal 3 segments elongate (Fig. 80:E,I); areola wide hexagonal and a little arched, area dentipara distinctly narrowed, apophysis strong, narrow, and toothlike, propodeal carinae distinct; 1st tergum moderately slender (Fig. 84:C), postpetiole gradually widened, about 1.5 time as wide as long; femur III moderately swollen (Fig. 79:D), ratio 0.24; Rs straight and slightly bowed (Fig. 82:A). *Other*: 3rd valvula long, about as long as hind tarsomeres; wings tinted.

REMARKS. This species is closely related to eastern *mucronatus*. The male of *taiganus* is distinguished by the orange abdomen, black scape, and coxae entirely black, without white apically, tibia III never with white dorsally. Orange forms of *mucronatus* have white on the scape and coxae. Female *taiganus* is distinguished by the orangish tiba III with black apically, never with white dorsal stripe, and its western distribution. See additional remarks under *texanus*.

RANGE. Alaska and British Columbia south and eastward to South Dakota and Colorado. Collections July and August.

MATERIAL EXAMINED. **Holotype**: male, Stone Mt. Pk., 3500 ft., BRITISH COLUMBIA, 23-vii-1973, H. and M. Townes (AEI). **Paratypes**: 33 M, 9 F (AEI, CAS, CNC, INF, USNM): ALASKA, BRITISH COLUMBIA, COLORADO, NEWFOUNDLAND, SOUTH DAKOTA, YUKON TERRITORY. Other material studied: 11 M (INF): ALASKA.

ETYMOLOGY. Taiga (Canadian coniferous forests) + -*anus* (adjectival suffix), referring to the habitat of the type series.

Endasys texanus (Cresson)
(Fig. 24:A-D; Map 14)

Phygadeuon texanus Cresson, 1872: 160; Dalla Torre, 1902: 695; Viereck, (1916) 1917: 334 and 336. **Holotype**, here determined.
Endasys (Endasys) texanus: Townes, 1944: 216; Townes and Townes, 1951: 240.
Endasys texanus: Carlson, 1979: 418.

MALE DIAGNOSIS. Large, 7-9 mm long; head rugulose with very dense, coarse pits; white scape, black clypeus; 3rd tyloid on flagellar segment 12; propodeum with strong carinae, apophysis very strong and toothlike; coxae black or orange, white apically, trochanters white ventrally, femur III and tibia III orange with black apices; 1st abdominal tergum moderately slender (Fig. 84:F), postpetiole longer than wide or squarish, dorsomedian carina strong; eastern half of United States and southern Canada.

MALE DESCRIPTION. *White*: Scape, tegula sometimes, coxae apically, trochanters ventrally, tibiae I and II dorsally, sometimes tibia III dorsally. *Yellow*: Femora I and II, tibiae I and II except latter dorsally. *Orange*: Flagellum often ventrally on basal half, usually coxae basally, femur III except apically, tibia III except apically and often dorsally, abdomen except parameres. *Black*: Clypeus, flagellum at least apical half, often tegula, often coxae basally, abdomen apically and often basally. *Punctation* (Fig. 86:F): Face, frons, and temple rugulose with very dense, coarse pits; propleurum slightly rugulose with very dense to sparse, coarse pits; mesopleurum impunctate and a little rugulose centrally, otherwise rugulose with dense, coarse pits; 1st and 2nd terga slightly mat and a little wrinkled. *Shape*: Clypeal margin not upturned (Fig. 78:A); flagellum moderately slender (Fig. 80:B), its segments longer than wide to apex; areola widely hexagonal, nearly quadrangular, area dentipara narrowed, apophysis very strong and toothlike, propodeal carinae strong and rugulose; 1st tergum moderately slender (Fig. 84:F), postpetiole longer than wide or square, dorsomedian carina strong, 2nd tergum regular or slightly widened; femur III moderately slender (Fig. 79:B), ratio 0.21-0.22; Rs straight, very slightly bowed (Fig. 82:A). *Other*: Wings more or less hyaline, 5th abdominal sternum complete medially.

FEMALE DIAGNOSIS. Large, 8-9.5 mm long; white flagellar segments 4-9 or 10; orange scape, black clypeus and mandible, leg III orange except tibia apically and tarsomeres blackish; head and thoracic pleura rugulose with very dense, coarse pits; propodeum with strong, toothlike apophysis; 1st abdominal tergum moderately slender (Fig. 84:C), postpetiole gradually widened, about twice as wide as long; 3rd valvula long, as long or longer than basal 4 hind tarsomeres.

FEMALE DESCRIPTION. *White*: Flagellar segments 4 to 9 or 10, tibiae I and II dorsally. *Yellow*: Legs I and II except tibiae dorsally and coxae basally; hind trochanters. *Orange*: Scape, tegula sometimes, leg III except trochanters and tibia apically, most of tarsomeres; abdomen entirely. *Black*: Flagellum except annulus, clypeus, mandible, tegula usually, hind tarsomeres more orangish black, tibia III apically. *Punctation* (Fig. 86:F): Head and thoracic pleura rugulose with very dense, coarse pits. *Shape*: Clypeal margin not upturned (Fig. 78:A); flagellum moderately slender (Fig. 80:B), distinctly narrowed apically; areola widely hexagonal, area dentipara narrowed, slightly elongate apically, apophysis strong and toothlike, propodeal carinae distinct; 1st tergum moderately slender (Fig. 84:F), postpetiole gradually widened, about twice as wide as long; femur III moderately swollen (Fig. 79:C), ratio 0.22-0.24; Rs straight, often slightly bowed (Fig. 82:A). *Other*: 3rd valvula long, as long or longer than basal 4 hind tarsomeres; wings hyaline or slightly tinted.

REMARKS. This species is closely related to *mucronatus*, *taiganus*, and Palearctic *cnemargus* (Gravenhorst). Male *texanus* differs from *mucronatus* by the orange abdomen and femur III and tibia III orange with black apically, tibia III never with white stripe dorsally. It differs from *taiganus* and *cnemargus* by the orange hind femur and tibia, and the coxae with white apically. Female *texanus* differs from all three by the orange scape, coxae, and hind femur.

RANGE. Eastern half of the United States and southern Canada from Minnesota to New Brunswick, southward to Texas, Louisiana, and Florida. Collections June through August.

MATERIAL EXAMINED. **Holotype**: male, Comal Co., TEXAS, 1872, Cresson (ANSP). **Paratypes**: (4 M unaccounted for) 2 M (ANSP, USNM): TEXAS. Other material studied: 65 M, 10 F (AEI, CNC, Dasch, FLDA, USNM): CONNECTICUT, FLORIDA, KENTUCKY, MAINE, MARYLAND, MICHIGAN, NEW BRUNSWICK, NEW HAMPSHIRE, NEW JERSEY, NEW YORK, NORTH CAROLINA, OHIO, PENNSYLVANIA, RHODE ISLAND, SOUTH CAROLINA, TENNESSEE, WEST VIRGINIA.

ETYMOLOGY. *Texanus*, referring to the type locality, Texas.

Endasys xanthopyrrhus Luhman, sp. n.
(Fig. 21:A-D; Map 21)

MALE DIAGNOSIS. Medium large, 6-7 mm long; scape yellow, clypeus yellowish without upturned margin, coxae I and II orangish with white apices, coxa III orange with some black basally, trochanters orangish and white, femur III and tibia III orange with black apex, abdomen weakly bicolored with blackish apically and laterally on terga 3-6, base and apex black; femur III moderately slender (Fig. 79:B), ratio 0.21-0.23; 1st abdominal tergum moderately stout (Fig. 84:E), postpetiole square to a little longer than wide; frons very densely punctate with finely distinct pits, surface slightly rugulose; Pacific Northwest.

MALE DESCRIPTION. *White*: Scape and mandible sometimes, coxae I and II apically, at least trochantelli ventrally. *Yellow*: Scape often more yellow-orange, usually clypeus and mandible, and tegula. *Orange*: Flagellum pale orange ventrally, often coxae I and II basally, coxa III except dorsally, 1st-trochanters usually more orangish or blackish, femur III and tibia III except apices, abdomen weakly bicolored with orange on 1st tergum apically, tergum 2 entirely, and variably on terga 3-6 basally. *Black*: Flagellum except ventrally, sometimes coxae I and II basally, often coxa III dorsally, femur III and tibia III apically, hind tarsomeres, abdomen basally and apically, and variably on terga 3-6 apically and laterally. *Punctation* (Fig. 86:C,E-F): Face and frons finely rugulose with very dense pits; temple with moderately dense, indistinct pits; propleurum with variably dense to sparse, distinct pits, surface a little rugulose; mesopleurum impunctate centrally, otherwise mostly moderately sparse with distinct pits, surface slightly rugulose on lower part; 1st and 2nd terga mat. *Shape*: Clypeus without upturned margin (Fig. 78:A); flagellum moderately slender (Fig. 80:B), its segments longer than wide on apical half, sometimes square; areola widely hexagonal, area dentipara narrowed, apophysis more or less distinct, propodeal carinae distinct; 1st tergum moderately stout (Fig. 84:E), postpetiole squarish or a little longer than wide, 2nd tergum wide; femur III moderately slender to moderately swollen, ratio 0.21-0.23; radial cell long, Rs straight (Fig. 82:A). *Other*: Wings more or less hyaline, 5th abdominal sternum complete medially, glumes dense (Fig. 88:A).

FEMALE DIAGNOSIS. Medium large, 6-7 mm long; clypeus orange, margin sharp and very slightly upturned (Fig. 78:B); flagellum orange, short and slender, narrowed apically (Fig. 80:E), basal 3 segments short; legs entirely orange, abdomen orange or slightly blackish apically on terga 3-apex; 3rd valvula moderately long, as long as basal 4 hind tarsomeres; area dentipara distinctly narrowed, apophysis moderately distinct; 1st tergum moderately stout (Fig. 84:B), postpetiole moderately widened, about 1.5 times as wide as long; radial cell long, Rs straight; frons very densely punctate with fine pits, surface finely rugulose; femur III moderately swollen (Fig. 79:C), ratio about 0.26.

FEMALE DESCRIPTION. *Orange*: Flagellum, scape, clypeus, mandible, tegula, legs, abdominal terga 1 and 2 entirely, terga 3-apex basally. *Black*: Apically on terga 3-apex; dark forms with more black on all the above except abdomen distinctly bicolored, distinctly black on terga 3-apex. *Punctation* (Fig. 86:B,D): Face rugulose with dense pits; frons with dense to very dense, finely distinct pits; temple with moderately dense, more or less indistinct pits; propleurum with dense, fine, more or less distinct pits; mesopleurum variable with more or less distinct pits, surface a little rugulose. *Shape*: Clypeus with sharp margin very slightly upturned (Fig. 78:B); flagellum short and slender, distinctly narrowed apically, basal 3 segments short (Fig. 80:E,H); areola broadly hexagonal, area dentipara distinctly narrowed, apophysis moderately distinct, propodeal carinae distinct; 1st tergum moderately stout (Fig. 84:E), postpetiole moderately widened, about 1.5 times as wide as long; femur III moderately swollen (Fig. 79:D), ratio 0.26; radial cell long, distinctly longer than 2nd-discoidal cell, Rs straight. *Other*: 3rd valvula moderately long, as long as basal 4 hind tarsomeres or longer; wings slightly tinted.

REMARKS. This species is closely related to *xanthostomus* from which males are distinguished by the yellow scape and tegula, and white on the coxae and trochanters. Females differ by the orange flagellum, not bicolored, basal 3 segments short, and the hind femur more swollen (ratio 0.26). It also appears related to Palearctic *amoenus* (Habermehl), from which both sexes differ by the distinctly bicolored abdomen and the clypeus not widened.

RANGE. Alaska and western Canada southward to Washington and Oregon. Collections in July.

MATERIAL EXAMINED. **Holotype**: male, Stone Mt. Pk., 3800+ ft., BRITISH COLUMBIA, 13-vii-1973, H. and M. Townes (AEI). **Paratypes**: 17 M, 6 F (AEI, Dasch): ALASKA, ALBERTA, BRITISH COLUMBIA, OREGON, WASHINGTON, YUKON TERRITORY.

ETYMOLOGY. (Greek) *Xantho-* (yellow) + *pyrrhus* (redish orange), referring to the yellow clypeus and the orange legs.

Endasys xanthostomus Luhman, sp. n.
(Fig. 40:A-D; Map 22)

MALE DIAGNOSIS. Medium large, 5-7 mm long; scape black, clypeus yellow, legs mostly orange, coxae sometimes black, sometimes apices of femur III and tibia III black, abdominal terga 2 and 3 mostly orange, terga 4-apex black; propodeum sloped with apophysis distinct and sharp, area dentipara narrowed; femur III slender (Fig.79:A), ratio 0.19-0.20; 1st abdominal tergum moderately stout (Fig. 84:E), postpetiole longer than wide; postpetiole and 2nd terga slightly mat; tyloids long and sharp, but not prominent; western Canada, Oregon, and Idaho.

MALE DESCRIPTION. *Yellow*: Clypeus and usually mandible. *Orange*: Scape orangish black, often coxae, trochanters usually, femora except femur III apically, tibiae except tibiae III dorso-apically, abdominal tergum 1 apically, terga 2 and 3 except variably apically and laterally; light forms with mostly orange legs except apices of femur II and tibia III. *Black*: Scape, tegula, mandible often, coxae sometimes, femur III apically, tibia III dorso-apically, hind tarsomeres, abdomen basally, terga 4-apex, variably on terga 2 and 3 apically and laterally. *Punctation* (Fig. 86:A-D): Face finely rugulose with very dense, fine pits; frons with mostly dense, fine pits; temple without distinct pits, hair sockets dense; propleurum indistinctly punctate to impunctate; mesopleurum impunctate centrally, otherwise with variably dense or sparse, more or less distinct, surface more or less rugulose; 1st and 2nd terga slightly mat. *Shape*: Clypeus convex without upturned margin (Fig. 78:A); flagellum moderately slender (Fig. 80:B), its segments longer than wide to apex; areola broadly to widely hexagonal, area dentipara distinctly narrowed and short, apophysis distinct and sharp, propodeal carinae distinct; 1st tergum moderately stout (Fig. 84:E), postpetiole longer than wide, 2nd tergum regular; femur III slender (Fig. 79:A), ratio 0.19-0.20; radial cell long, about 1.5 times as long as 2nd-discoidal cell, Rs straight. *Other*: Sometimes short 3rd tyloid on flagellar segment 9, wings more or less hyaline, 5th abdominal sternum complete medially, and glumes moderately sparse (Fig. 88:B).

FEMALE DIAGNOSIS. Medium large, about 6 mm long; legs and abdomen orange, flagellum slightly bicolored and narrowed apically (Fig. 80:E); propodeum short, area dentipara distinctly narrowed, apophysis distinct; femur III moderately slender (Fig. 79:B), ratio 0.23; 3rd valvula long, about as long as basal 5 tarsomeres; radial cell elongate, about 1.5 times as long as 2nd-discoidal cell, Rs straight.

FEMALE DESCRIPTION. *Orange*: Flagellum except darker on apical half, scape, clypeus, mandible, tegula, legs, abdomen. *Punctation* (Fig. 86:B-C): Face finely rugulose with very dense, fine pits; frons with dense, fine, more or less distinct pits; temple with dense, indistinct pits; propleurum with dense, fine pits; mesopleurum with variably sparse to dense punctation, fine, more or less distinct pits. *Shape*: Clypeus convex, margin weakly upturned (Fig. 78:B); flagellum slender, narrowed apically, basal 3 segments elongate (Fig. 80:E,I); areola broadly hexagonal, area dentipara narrowed, apophysis distinct and vertically projecting, propodeal carinae weak; 1st tergum moderately stout (Fig. 84:B), spiracle near midpoint, postpetiole moderately widened, about 1.5 times as wide as long; femur III moderately slender (Fig. 79:B), ratio 0.22-0.23; radial cell elongate, about 1.5 times as long as 2nd-discoidal cell, Rs straight. *Other*: 3rd valvula long, about as long as basal 5 tarsomeres; wings hyaline to slightly tinted.

REMARKS. See discussion under *xanthopyrrhus*.

RANGE. Western Canada, Oregon, and Idaho. Collections May through August, mostly June and July.

MATERIAL EXAMINED. **Holotype:** male, Idaho City, IDAHO, 13-vi-1978, H. and M. Townes (AEI). **Paratypes:** 6 M, 4 F (AEI, CNC): ONTARIO, OREGON, IDAHO. Other material studied: 10 M, 2 F (AEI, CAS, Dasch, OSU): BRITISH COLUMBIA, CALIFORNIA, OREGON, YUKON TERRITORY.

ETYMOLOGY. (Greek) *Xantho-* (yellow) + *stomus* (mouth), referring to the yellow clypeus.

SUBCLAVATUS GROUP

DIAGNOSIS. Clypeal margin not or weakly upturned, propodeal carinae strong, apophysis distinct, male with at least front and middle coxae and trochanters partly or mostly white, scape and clypeus usually yellow or white, female with flagellum usually bicolored, basal 3 segments usually short, front and middle coxae and trochanters yellow, scape yellow, clypeus yellow, orange, or sometimes blackish; mostly eastern North America.

This group appears most closely related to the Texanus Group. I have included 16 species; *tyloidiphorus* is included with reservations. The group is characterized as follows: sexual dimorphism and dichromatism distinct, clypeus convex to flat, margin not or weakly upturned; male face width about equal to height of face plus clypeus; female flagellum usually linear to apex, basal 3 segments short, flagellum without distinct annulus, usually bicolored; male flagellum typically with 2 tyloids; punctation of head unevenly dense to moderately dense, usually some rugulosity on face and frons, especially females, male mesopleural plate mostly smooth and impunctate centrally, some rugulosity on lower part; males with white or pale yellow on scape, clypeus, tegula, at least front and middle coxae apically, trochanters, and often on front and middle tibiae dorsally; females with foregoing areas orangish or yellowish, never distinctly white; abdomen generally orange, males often with 1st and 2nd terga black, *praerotundiceps* male entirely black; propodeum more or less sloped in lateral view, more so in males, carinae moderately strong in males, distinct but finer in females, apophysis distinct, usually shorter and wider in females, area dentipara narrowed or regular, often a little elongate apicolaterally, areola widely to broadly hexagonal in males, broadly so in females, often a little elongate anteriorly; 1st abdominal tergum of male moderately slender to moderately stout, postpetiole square or elongate, dorsomedian carina distinct beyond spiracle; 1st tergum of female moderately stout to stout; hind femur generally moderately swollen; 5th abdominal sternum of male complete or membranous medially, 6th sternum always complete.

Species of this group are most common in the eastern half of the U.S and southern Canada; however some species are very common in the West and Pacific Northwest. Reared hosts are diprionid and tenthredinid cocoons.

This group corresponds to the Palearctic Testaceus Group. The following Nearctic species are included: *albior, aurarius, aurigena, brevicornis, chrysoleptus, hesperus, inflatus* (Provancher), *michiganensis, nemati, paludicola* (Brues), *patulus* (Viereck), *praerotundiceps, pubescens* (Prov.), *rotundiceps* (Prov.), *subclavatus* (Say), and *tyloidiphorus.*

Endasys albior Luhman, sp. n.
(Fig. 62:A-D; Map 1)

MALE DIAGNOSIS. Medium size, 5-6.5 mm long; white scape, clypeus, coxae I and II apically, trochanters; 1st abdominal tergum moderately slender (Fig. 84:F), postpetiole a little longer than wide, 2nd tergum regular or slightly narrowed; abdomen generally orangish with black basally, apically, and laterally and apically on terga 2-6; femur III slender (Fig. 79:A), ratio 0.20, usually orangish basally and black apically; propodeum slightly sloped, areola broadly hexagonal, elongate posteriorly, area dentipara nearly regular, apophysis distinct, sometimes weak; flagellum moderately slender (Fig. 80:B), segments a little longer than wide on apical half; Colorado.

MALE DESCRIPTION. *White*: Scape, clypeus usually, mandible, coxae I and II apically, trochanters except hind 1st-trochanter more blackish. *Yellow*: Femora I and II, tibiae I and II. *Orange*: Femur III mostly basally and ventrally, tibia III apically, variably on abdomen, usually 1st tergum apically, and terga 2-6 apically and laterally. *Black*: Flagellum, tegula, coxae except coxae I and II apically, hind 1st-trochanters blackish, femur III at least apically, tibia III apically, hind tarsomeres, abdomen at least basally and apically, and terga 2-6 apically and laterally; dark forms with mostly black femur III and abdomen. *Punctation* (Fig. 86:A-C): Face finely granular or finely rugulose with dense pits, surface shiny; frons with dense, very fine pits; temple with moderately dense, indistinct pits; propleurum with dense to sparse, very fine, indistinct pits, surface smooth and shiny; mesopleurum impunctate centrally, otherwise with mostly sparse, very fine pits, surface smooth and shiny; 1st and 2nd terga smooth and shiny. *Shape*: Clypeal margin slightly upturned (Fig. 78:B); flagellum moderately slender (Fig. 80:B), segments a little longer than wide on apical half; areola elongate hexagonal, area dentipara regular, apophysis moderately distinct, propodeum slightly sloped, carinae moderately distinct; 1st tergum more or less slender, postpetiole longer than wide, 2nd tergum regular or a little narrowed; femur III slender (Fig. 79:A), ratio 0.20; radial cell long, Rs straight. *Other*: Wings more or less hyaline, 5th abdominal sternum complete medially, and glumes dense.

FEMALE DIAGNOSIS. Medium small, 5.5 mm long; orange, slender, flagellum, narrowed apically (Fig. 80:E), basal 3 segments a little elongate; clypeus black, margin slightly upturned (Fig. 78:B); punctation of frons dense with fine, more or less distinct pits; abdomen mostly orange on basal half, blackish on apical half; 1st tergum moderately stout (Fig. 84:B), postpetiole moderately widened, about 1.5 times as wide as long; 3rd valvula moderately long, as long as basal 4 hind tarsomeres; yellow-orange coxae, trochanters, femora I and II, and all tibiae; blackish coxae I and II and femur III.

FEMALE DESCRIPTION (based on one specimen). *Orange*: Flagellum, scape, mandible, tegula, legs except middle and hind coxae and hind femur more blackish orange, and abdomen orangish on basal half. *Black*: Clypeus, middle and hind coxae more blackish orange, hind femur blackish orange, and abdomen more orangish black on apical half. *Punctation* (Fig. 86:B-C): Face finely rugulose with dense pits; frons and propleurum with dense, fine, more or less distinct pits, propleurum smooth and shiny; temple with moderately sparse, indistinct pits; mesopleurum mostly smooth and shiny with sparse, very fine pits. *Shape*: Clypeal margin slightly upturned (Fig. 78:B); flagellum slender, narrowed apically, basally 3 segments a little elongate (Fig. 80:E,H); areola broadly hexagonal, area dentipara narrowed, a little elongate apicolaterally (Fig. 83:C), apophysis moderately strong, projecting at an angle, propodeal carinae fine but distinct; 1st tergum moderately stout (Fig. 84:B), postpetiole moderately widened, more than 1.5 times as wide as long; femur III moderately swollen (Fig. 79:D), ratio 0.25; radial cell slightly curved. *Other*: 3rd valvula moderately long, as long as basal 4 hind tarsomeres; wings tinted.

REMARKS. The female described is part of a series of 3 males having the same locality and dates. The correspondence of color pattern, punctation, propodeal shapes, and the shape of the clypeus are also very good. For affinities of the species, see remarks under *paludicola*.

RANGE. Colorado. Collections July and August.

MATERIAL EXAMINED. **Holotype**: male, Poudre L., Rocky Mt. Natl. Pk., COLORADO, 12-VIII-1948, H., G. and D. Townes (AEI). Paratypes: 55 M, 1 F (AEI, Dasch): COLORADO. Other material studied: 30 M (AEI, Dasch): COLORADO.

ETYMOLOGY. *Alb-* (white) + *ior* (-er), referring to the white of the head and legs.

Endasys aurarius Luhman, sp. n.
(Fig. 13:A-D; Map 12)

MALE DIAGNOSIS. Large, 7-9 mm long; orange thorax and propodeum, leg III except white trochanters, and abdomen; black head capsule, white scape, mandible, and clypeus; frons dense with distinct, sometimes coarse pits, propleurum densely punctate to impunctate; areola broadly to widely hexagonal, area dentipara narrowed, distinct apophysis; 1st abdominal tergum long and slender, postpetiole longer than wide, dorsomedian carina weak, postpetiole appearing smooth dorsally; 2nd tergum widened; eastern half of the United States and southern Canada.

MALE DESCRIPTION. *White*: Scape, clypeus, mandible, tegula, coxae I and II except basally, trochanters, tibiae I and II dorsally. *Yellow*: coxae I and II basally, femora I and II , tibiae I and II except latter dorsally, sometimes hind tarsomeres more pale yellowish. *Orange*: Flagellum at least ventrally; most of thorax and propodeum, variably on sterna, wing base, and along sutures and carinae; leg III except often tibia III blackish apically, and sometimes tarsomeres more blackish or yellowish; abdomen including apex. *Black*: Flagellum more blackish dorsally and apically; head capsule except clypeus; often thorax and propodeum on carinae, wing base, along sutures, sulci, and ventrally. *Punctation* (Fig. 86:D-F): Face rugulose with very dense pits; frons with dense, distinct, often coarse pits; temple with dense, more or less distinct pits; propleurum densely punctate to impunctate, pits more or less distinct, or mostly sparsely punctate; mesopleurum impunctate and slightly rugulose centrally, otherwise punctation variable with distinct pits, surface with rugulosity; postpetiole and 2nd tergum smooth and shiny. *Shape*: Clypeal margin not or very slightly upturned; flagellum moderately slender (Fig. 80:B), segments longer than wide to apex; areola broadly to widely hexagonal, apophysis distinct, propodeal carinae distinct; 1st tergum long and slender, postpetiole squarish or longer than wide, sides very slightly diverging apically, dorsomedian carina weak, often apparently without carinae, surface smooth; 2nd tergum widened; femur III moderately swollen (Fig. 79:C), ratio 0.23-0.24; Rs very slightly curved, radial cell a little widened, nearly as high as 2nd-discoidal cell. *Other*: Wings tinted. 5th abdominal sternum membranous medially, glumes dense, and tyloids (2) moderately long.

FEMALE DIAGNOSIS. Medium to large, 5.5-7 mm long; mostly orange thorax and propodeum, abdomen, and leg III except trochanters yellow; flagellum bicolored, more or less linear to apex (Fig. 80:F), apical 3 segments very slightly narrowed, basal 3 segments short; 3rd valvula moderately long, about as long as hind tarsomeres; orange clypeus with sharp margin, very slightly upturned (Fig. 78:B); punctation of head and thoracic pleura generally dense to very dense and coarse, surface rugulose; 1st abdominal tergum moderately slender (Fig. 84:F), postpetiole moderately widened to about 1.5 times as wide as long, sides nearly parallel; area dentipara distinctly narrowed, apophysis distinct; femur III moderately swollen (Fig. 79:D), ratio 0.28-0.29; radial cell long, Rs straight.

FEMALE DESCRIPTION. *Yellow*: Flagellum on basal half, scape, tegula, legs I and II, and hind trochanters. *Orange*: Clypeus, mandible, usually entire thorax and propodeum except prosternum, leg III except trochanters, abdomen. *Black*: Flagellum on apical half, head capsule except clypeus and mandible, and usually propleurum, sometimes wing base and thorax ventrally. *Punctation* (Fig. 86:E-F): Face, frons, propleurum, and mesopleurum rugulose with very dense pits; temple slightly rugulose with dense pits. *Shape*: Clypeus with sharp margin very slightly upturned; flagellum moderately stout, more or less linear to apex, apical 3 segments very slightly narrowed, basal 3 short; areola broadly hexagonal, area dentipara narrowed, apophysis distinct, propodeal carinae distinct; 1st tergum moderately slender (Fig. 84:C), postpetiole moderately widened, about 1.5 times as wide as long, sides nearly parallel; femur III moderately swollen (Fig. 79:C), ratio 0.28-0.29; radial cell long, Rs straight. *Other*: 3rd valvula moderately long, about as long as hind tarsomeres; wings slightly tinted.

REMARKS. This species is closely related to *patulus* and *subclavatus*. Both males and females are distinguished by having the thorax and propodeum orange, punctation of the head and propleurum coarser, and (females) longer ovipositor, about as long as hind 5 tarsomeres.

Both orange and black forms are found in collections from the same rearings (USNM and FLDA), but orange forms are more common. The black forms resemble *patulus* (Viereck), but have a stouter flagellum, clypeal margin not upturned, coarse punctation of the face and frons, and more swollen femur III. Female *aurarius* in addition has a distinctly long 3rd valvula, and moderately widened postpetiole.

Sometimes orange forms of *patulus* occur, but both males and females can be distinguished from *aurarius* by the same characters listed above. Orange forms lack the black melanin, as they are nearly transparent, muscle attachments being visible through the integument. This phenomenon of black and orange forms of the same species is found in other Gelinae and in Ichneumoninae. It may have some phylogenetic significance, or simply an indication that the lack of black melanin of the head and thorax is determined by a single, perhaps recessive, gene.

RANGE. Eastern half of the United States and southern Canada, more common in collections from the southeastern U.S., especially North Carolina to Florida. Collections April through July.

MATERIAL EXAMINED. **Holotype**: male, Cleveland, SOUTH CAROLINA, 27-v-1971, G. Townes (AEI). **Paratypes**: 33M, 5 F (AEI, CNC, FLDA): CONNECTICUT, GEORGIA, KENTUCKY, MICHIGAN, OHIO, ONTARIO, SOUTH CAROLINA. Other material studied: 23 M, 9 F (AEI, FLDA, USNM): FLORIDA, NEW YORK, NORTH CAROLINA, OHIO, VIRGINIA.

ETYMOLOGY. *Aur-* (gold) + *-arius* (-en, adjectival suffix), referring to the golden orange of the thorax and propodeum.

Endasys aurigena Luhman, sp. n.
(Fig. 54:A-D; Map 4)

MALE DIAGNOSIS. Medium, 6-6.5 mm long; coxa III mostly orange with black basally and white ventro-apically, femur III and tibia III orange with black apices, hind tarsomeres black, trochanters white, coxae I and II white with yellow or orangish basally, abdomen orange except basally and apically; 1st abdominal tergum moderately stout (Fig. 84:E), postpetiole square to wide, 2nd tergum wide; areola widely hexagonal (Fig. 85:D), area dentipara strongly narrowed, apophysis strong, propodeal carinae strong; flagellum moderately stout (Fig. 80:C), its segments square on apical half; eastern half of the United States and southern Canada.

MALE DESCRIPTION. *White*: Scape, sometimes clypeus, coxae I and II except basally, coxa III ventroapically, trochanters, tibiae I and II dorsally. *Yellow*: Clypeus usually, coxae I and II basally (sometimes more orangish), femora I and II, and tibiae I and II except dorsally. *Orange*: Hind coxa III except ventroapically, femur III and tibia III except apically, sometimes hind tarsomeres blackish orange, abdomen except basally and apically. *Black*: Femur III and tibia III apically, usually hind tarsomeres, abdomen basally and apically. *Punctation* (Fig. 86:B-C,E-F): Face finely rugulose; frons with very dense, distinct pits, a little rugulose; temple with dense, indistinct pits; propleurum densely punctate to impunctate, very fine pits; mesopleurum impunctate centrally, otherwise with sparse, distinct pits, a little rugulose on lower part. *Shape*: Clypeal margin not upturned (Fig. 78:A); flagellum moderately stout (Fig. 80:C), its segments square on apical half; areola widely hexagonal (Fig. 85:D), area dentipara strongly narrowed, apophysis strong and toothlike, propodeal carinae strong; 1st abdominal tergum moderately stout (Fig. 84:E), postpetiole square to wide, 2nd tergum wide; femur III moderately swollen (Fig.

79:C), ratio 0.24. *Other*: Tyloids (2) moderately short (Fig. 87:C) and prominent, wings hyaline, 5th abdominal sternum membranous medially, and glumes moderately dense (Fig. 88:D).

FEMALE DIAGNOSIS. Medium, 6 mm long; hind coxa blackish with yellow ventroapically, femur III and tibia orange with black apices; abdomen orange except 1st tergum black laterally; yellow scape and basal half of flagellum, segments 5-8 paler; flagellum more or less linear to apex, basal 3 segments short; orange clypeus weakly upturned (Fig. 78:B); 1st tergum stout (Fig. 84:A); 3rd valvula moderately long, about as long as basal 4 hind tarsomeres; frons with dense, distinct pits, a little rugulose, temple with dense, indistinct pits; area dentipara narrowed, elongate apico-laterally (Fig. 83:C), apophysis and propodeal carinae distinct.

FEMALE DESCRIPTION. *Yellow*: Scape, flagellum on basal half, flagellar segments 5-8 pale yellow, legs I and II except femur II blackish posteriorly, coxa III ventro-apically, and hind trochanters. *Orange*: Clypeus, mandible, tegula, femur III and tibia III except apically, hind tarsomeres blackish orange, and abdomen except 1st tergum laterally. *Black*: Coxa III basally, femur III and tibia III apically, and 1st tergum laterally. *Punctation*: Face finely granular; frons with dense, distinct pits, a little rugulose; temple with moderately dense, indistinct pits; propleurum with dense, fine pits; mesopleurum rugulose with sparse, more or less distinct pits. *Shape*: Clypeal margin weakly upturned (Fig. 78:B); flagellum more or less linear to apex, basal 2 segments short (Fig. 80:F,H); areola broadly hexagonal, area dentipara narrow, a little elongate apico-laterally (Fig. 83:C), apophysis distinct, propodeal carinae distinct; 1st tergum stout (Fig. 84:A), postpetiole twice as wide as long; femur III swollen (Fig. 79:E), ratio 0.28; Rs more or less straight. *Other*: 3rd valvula moderately long, about as long as basal 4 hind tarsomeres; wings slightly tinted.

REMARKS. This species is probably more closely related to *subclavatus*. Males are distinguishable mainly by the black hind tarsomeres, never whitish, mostly orange coxa II, and the stouter 1st abdominal tergum, postpetiole nearly square. Females are distinguished by the distinctly black apices of the femur III and tibia III, 1st abdominal tergum partly black, femur III swollen (ratio about 0.3), the finer punctation of the head, and the stouter postpetiole.

RANGE. Eastern half of United States and southern Canada. Collections March through November, mostly June through September.

MATERIAL EXAMINED. **Holotype**: male, High Pt. St. Pk., NEW JERSEY, 2-vi-1973, R. Reardon (AEI). **Paratypes**: 67 M (AEI, CNC, Dasch, FLDA, USNM): ARKANSAS, FLORIDA, KANSAS, MARYLAND, MICHIGAN, NEBRASKA, NEW BRUNSWICK, NEW HAMPSHIRE, NEW JERSEY, NEW YORK, NORTH CAROLINA, OHIO, ONTARIO, PENNSYLVANIA, SASKATCHEWAN, WEST VIRGINIA. Other material studied: 92 M, 2 F (Dasch, FLDA, UCD, UM): FLORIDA, MARYLAND, MICHIGAN, MINNESOTA, OHIO, PENNSYLVANIA, WASHINGTON D.C.

ETYMOLOGY. *Auri-* (gold) + *gena* (begotten; gender m. and f.), an epithet of Perseus as the begotten of Danae, the Golden; name here refers to the concept of the species born from "golden" *subclavatus*.

Endasys brevicornis Luhman, sp. n.
(Fig. 18:A-D; Map 19)

MALE DIAGNOSIS. Small, 4-5 mm long; flagellum short and stout (Fig. 80:D), its segments squarish or widened on apical half; white on coxae apically, trochanters, tegula, scape, and clypeus, sometimes latter 3 more yellowish; coxa III blackish and orange basally, femur III black at least apically, basally orange, tibia III orange with black apex; abdomen weakly bicolored, black basally, on tergum 3 apically and laterally, and on terga 4-apex; face finely rugulose, frons with moderately dense to moderately sparse punctation, very fine, more or less distinct pits; propodeum with distinct carinae, area dentipara more or less narrowed, apophysis moderately strong; femur III swollen (Fig. 79:D), ratio 0.25; eastern half of United States and southern Canada.

MALE DESCRIPTION. *White*: Scape usually, clypeus, mandible, tegula, coxae apically, trochanters. *Yellow*: Flagellum often more yellowish ventrally; sometimes scape, clypeus, mandible more yellowish; coxae I and II except often more blackish basally; femora I and II, tibiae I and II. *Orange*: Coxa III except apically and often basally, femur III at least basally, tibia III except apically, abdomen on 1st tergum apically, 2nd tergum entirely, and 3rd tergum apically and laterally. *Black*: Flagellum except often yellowish ventrally, coxae I and II often basally, coxa III often basally, femur III at least apically, tibia III apically, abdomen basally, 3rd tergum apically and laterally, terga 4-apex. *Punctation* (Fig. 86:B-C): Face finely rugulose with dense pits; frons with moderately dense to moderately sparse punctation, very fine, more or less distinct pits; temple with moderately sparse, indistinct pits; propleurum sparsely punctate to impunctate, fine, more or less distinct pits; mesopleurum impunctate centrally, otherwise variably punctate with more or less distinct pits, surface slightly rugulose; 1st and 2nd terga often very slightly mat. *Shape*: Clypeus with sharp edge, margin more or less upturned (Fig. 78:B); flagellum short and stout (Fig. 80:D), its segments squarish to widened on apical half, basal half squarish to a little longer than wide; areola broadly hexagonal, area dentipara a little narrowed, apophysis moderately distinct, propodeal carinae distinct; 1st tergum moderately stout (Fig. 84:E), postpetiole squarish, 2nd tergum wide; femur III swollen, ratio 0.25; Rs very slightly curved. *Other*: wings more or less hyaline; 5th abdominal sternum more or less membranous medially, sometimes complete apically (Fig. 81:B); glumes moderately dense (Fig. 88:B).

FEMALE DIAGNOSIS. Small, 3-4.5 mm long; face finely rugulose with dense pits; clypeal margin sharp, more or less upturned (Fig. 78:C); coxae II and III blackish basally and yellow apically, coxa I mostly yellow, trochanters yellow, femur III blackish to yellowish black, tibia III yellowish with black apex; abdomen weakly bicolored, blackish basally, tergum 3 apically and laterally, and terga 4-apex; 3rd valvula moderately long, about as long as basal 4 hind tarsomeres; femur III swollen (Fig. 79:E), ratio about 0.29; 1st abdominal tergum moderately stout (Fig. 84:E), postpetiole more abruptly widened, sides nearly parallel.

FEMALE DESCRIPTION. *Yellow*: Flagellum basal half, scape, tegula, legs I and II, coxa III apically, trochanters, femur III at least basally, tibia III except apically, hind tarsomeres 1-4 except apically. *Pale Orange*: Clypeus blackish orange, mandible, and abdomen on tergum 1 apically, tergum 2 except very slightly laterally, and tergum 3 except at least apically and laterally. *Black*: Flagellum apical half, coxa III basally, most of femur III more yellowish black except basally, tibia III apically, apically on hind tarsomeres 1-4, abdomen basally, tergum 3 at least apically and laterally, terga 4-apex. *Punctation* (Fig. 86:B): Face finely rugulose with dense pits; frons with moderately sparse, finely distinct pits; temple with variably sparse, very fine, more or less distinct pits; propleurum with moderately sparse, fine, more or less distinct pits, surface smooth and shiny; mesopleurum with mostly sparse, fine, more or less distinct pits, more rugulose on lower part. *Shape*: Clypeal margin sharp and more or less upturned (Fig. 78:C); flagellum short and stout, appearing slightly clavate, apical half linear, basal 3 segments short (Fig. 80:F,H); areola broadly hexagonal, area dentipara narrowed, apophysis distinct, propodeal carinae weak; 1st tergum moderately stout (Fig. 84:B), postpetiole more abruptly widened, about 1.5 times as wide as long, sides nearly parallel; femur III swollen (Fig. 79:E), ratio about 0.29; radial cell short, Rs slightly curved (Fig. 82:D). *Other*: 3rd valvula moderately long, about as long as basal 4 hind tarsomeres; wings tinted.

REMARKS. This species may be more closely related to *michiganensis*. The male is distinguished by the short, stout flagellum, coxae yellowish basally, white apically, white clypeus, and femur III mostly yellowish orange with black apically. Females are distinguished by the orangish clypeus, yellow coxae, and clypeal margin sharp and upturned. See additional discussion under *nemati*.

HOST. Tenthredinidae: Nematinae: *Nematus salicisodoratus* Dyar, on willow (USNM).

RANGE. Eastern half of the United States and southern Canada, from Minnesota eastward to Ontario and New Brunswick, southward to Nebraska to South Carolina. Collections May through September; mostly July and August in the North, earlier and later in the South.

MATERIAL EXAMINED. **Holotype**: male, Big Fork, MINNESOTA, 11-vi-1972, H. and M. Townes (AEI). **Paratypes**: 104 M, 13 F (AEI, CNC, Dasch, UM, Yu): CONNECTICUT, ILLINOIS, MAINE, MARYLAND, MICHIGAN, MINNESOTA, MISSOURI, NEW BRUNSWICK, NEW JERSEY, NEW YORK, NORTH CAROLINA, OHIO, ONTARIO, QUEBEC, SOUTH CAROLINA, TENNESSEE, WEST VIRGINIA. Other material studied: 18M, 1 F (AEI, CNC, Dasch, UKS): MICHIGAN, NEBRASKA, NEW YORK, QUEBEC, SOUTH DAKOTA.

ETYMOLOGY. *Brevi-* (short) + *cornis* (horn) referring to the flagellum.

Endasys chrysoleptus Luhman, sp. n.
(Fig. 51:A-D; Map 7)

MALE DIAGNOSIS. Medium, 5.5-7 mm long; 1st abdominal tergum long and slender (Fig. 84:G), postpetiole longer than wide; femur III slender (Fig. 79:A), ratio 0.19-0.21; flagellum long and slender (Fig. 80:A), segments elongate to apex; white coxae I and II, coxa III apically, and trochanters; coxa III blackish and orange basally, femur III and tibia III orange with black apices, and abdomen orange with black basally and apically, and variably on 2nd tergum; propodeum short with strong carinae, areola widely hexagonal, area dentipara distinctly narrowed, apophysis distinct and vertically projecting; face more or less rugulose, frons with moderately dense to moderately sparse, distinct pits; southern Canada and eastern United States.

MALE DESCRIPTION. *White*: Tegula, most of coxae I and II, coxa III apically, and trochanters. *Yellow*: Scape, clypeus, mandible, coxae I and II at bases, femora I and II, tibiae I and II except latter pale yellow dorsally. *Orange*: Flagellum ventrally, hind coxa variably basally, femur III, tibia III except apices, hind tarsomeres variably, abdomen except usually basally and apically, variably on 2nd tergum. *Black*: Flagellum except ventrally, hind coxa variably at base, femur III and tibia III apically, hind tarsomeres variably, and abdomen basally and apically, variably on 2nd tergum. *Punctation* (Fig. 86:A-C,E): Face a little rugulose with dense pits; frons with moderately dense to moderately sparse punctation; temple with moderately sparse to moderately dense, indistinct pits; propleurum densely punctate to impunctate, pits fine, surface slightly rugulose; mesopleurum impunctate and a little rugulose centrally, otherwise with mostly sparse, distinct pits, surface rugulose; metapleurum distinctly rugulose; 1st and 2nd terga mostly smooth and shiny. *Shape*: Clypeal margin not upturned (Fig 78:A); face narrow, less than twice as wide as long; flagellum long and slender (Fig. 80:A) to moderately slender, its segments longer than wide to apex; areola widely hexagonal, area dentipara distinctly narrowed, apophysis strong and vertically projecting, propodeal carinae strong; 1st tergum long and slender (Fig. 84:G), postpetiole much longer than wide, 2nd tergum regular to slightly narrowed; femur III slender to moderately slender, ratio 0.19-0.21; Rs straight. *Other*: Wings slightly tinted, 5th abdominal sternum complete medially, and glumes dense.

FEMALE DIAGNOSIS. Medium large, 6-7 mm long; 3rd valvula long, about as long as hind tarsomeres; femur III slender to moderately slender (Fig. 79:A), ratio about 0.21; petiole long and slender, postpetiole gradually widened to about 1.5 times as wide as long; area dentipara narrowed, apophysis distinct and vertically projecting; face and frons with punctation dense and coarse; legs and abdomen mostly orange, blackish on femur III and tibia III apically, and sometimes on coxae II and III; flagellum bicolored, stout, linear to apex, basal 3 segments elongate (Fig. 80:F,I); radial cell long, sometimes apical 0.25 of Rs a little curved.

FEMALE DESCRIPTION. *Orange*: Flagellum basal half, scape, mandible, tegula, legs except usually apices of femur III and tibia III, sometimes coxae II and III blackish basally, abdomen. *Black*: Flagellum apical half, clypeus, femur III and tibia III apically orangish black. *Punctation* (Fig. 86:B,E): Face a little rugulose with dense, coarse pits; frons slightly rugulose with dense, distinct pits; temple with moderately dense, indistinct pits; propleurum with mostly dense, very fine, more or less distinct pits; mesopleurum with mostly sparse, very fine, more or less distinct pits, surface with rugulosity. *Shape*: Clypeal margin sharp but not upturned (Fig. 78:B); flagellum stout, linear to apex, basal 3 segments elongate (Fig. 80:F,I); areola broadly hexagonal, area dentipara distinctly narrowed, apophysis strong and vertically projecting, propodeal carinae weak or moderately distinct; petiole long and slender, postpetiole gradually widened to about 1.5

times as wide as long; femur III slender (Fig. 79:B), ratio about 0.21; Rs straight (Fig. 82:B). *Other*: 3rd valvula long, about as long as hind tarsomeres; wings tinted.

REMARKS. Males of this species are distinguished by the sharp clypeal margin, slender flagellum, femur III, and 1st abdominal tergum. It may be related to *rotundiceps*. Females are distinguished by the long 3rd valvula, about as long as the hind 5 tarsomeres, and the slender femur III (ratio 0.21).

RANGE. Southern Canada and eastern United States from Alberta to Newfoundland southward to West Virginia and North Carolina. Collections June through September.

MATERIAL EXAMINED. **Holotype**: male, Pisgah Mt., 4800 ft., NORTH CAROLINA, 2-ix-1950, H. and D. Townes (AEI). **Paratypes**: 81 M, 6 F (AEI, Dasch): MARYLAND, MICHIGAN, NEW YORK, NORTH CAROLINA, WEST VIRGINIA. Other material studied: 38 M (AEI, Dasch, USNM): ALBERTA, MARYLAND, MAINE, NEWFOUNDLAND, NEW HAMPSHIRE, NEW YORK, NORTH CAROLINA, OHIO, ONTARIO, PENNSYLVANIA, VERMONT, WEST VIRGINIA.

ETYMOLOGY. (Greek) *Chryso-* (gold) + *leptus* (slender), referring to the slenderness and color of the legs and abdomen.

Endasys hesperus Luhman, sp. n.
(Fig. 1:A-D; Map 17)

MALE DIAGNOSIS. Medium large, 6-7.5 mm long; face appearing finely granular, frons evenly dense with fine pits, yellow clypeus with sharp margin slightly upturned (Fig. 78:B), 1st abdominal tergum moderately slender (Fig. 84:F) with narrow postpetiole, pitting of postpetiole and 2nd tergum fine but distinct, surface slightly mat; coxae I and II white apically, trochanters white, femur III orange, tibia III orange with black apically, hind tarsomeres black, abdomen orange except apically and often basally; femur III moderately slender (Fig. 79:B), ratio about 0.20; wings darkened; Pacific Northwest.

MALE DESCRIPTION. *White*: Coxae I and II apically, trochanters except trochanter III often more yellowish. *Yellow*: Scape, clypeus, mandible, tegula, coxa I basally, femora I and II, tibiae I and II except dorsally pale yellow. *Orange*: Flagellum often ventrally, coxa II except apically, coxa III, femur III, tibia III except apically, abdomen except apically and sometimes basally. Sometimes scape, clypeus, and coxae yellowish orange. *Black*: Flagellum except often ventrally, tibia III apically, hind tarsomeres. Small forms often with femur III black apically. *Punctation* (Fig. 86:B-D): Face dense, appearing finely granular; frons with evenly dense, fine pits; temple with evenly dense, indistinct pits; propleurum densely punctate to impunctate, very fine, more or less indistinct pits, surface smooth and shiny; mesopleurum impunctate centrally, otherwise with mostly sparse, fine pits, surface smooth and shiny; postpetiole and 2nd tergum with sparse to dense, fine but distinct pits, very slightly mat. *Shape*: clypeus with sharp margin more or less upturned (Fig. 78:B); flagellum moderately slender (Fig. 80:B), its segments longer than wide on apical half; areola broadly hexagonal; area dentipara narrow to regular, a little elongate apicolaterally, apophysis distinct; 1st abdominal tergum slender to moderately slender (Fig.

84:F), postpetiole usually longer than wide; 2nd tergum regular; femur III moderately slender to moderately stout, ratio 0.20-0.23. *Other*: Wings darkened, 5th abdominal sternum complete apically, tyloids (2) long and thin (Fig. 87:A), and glumes dense to moderately dense.

FEMALE DIAGNOSIS. Medium large, about 6-7 mm long; flagellum orange, slender, and distinctly narrowed apically (Fig. 80:E); orange clypeus with slightly upturned margin; orange legs and abdomen; punctation of head and thoracic pleura generally fine, more or less distinct pits, surfaces mostly smooth and shiny; area dentipara distinctly narrowed, apophysis distinct; 1st abdominal tergum moderately stout, postpetiole widely expanded (Fig. 84:A), nearly twice as wide as long, sides parallel; 3rd valvula moderately short, about as long as basal 3 hind tarsomeres; femur III moderately swollen, ratio about 0.28; wings more or less darkened.

FEMALE DESCRIPTION. *Orange*: Flagellum, scape, clypeus except sometimes more blackish, tegula, legs except legs I and II yellowish orange, abdomen. *Black*: Clypeus sometimes more blackish. *Punctation*: (Fig. 86:B,D) Face appearing finely granular or finely rugulose; frons with evenly dense, more or less fine pits; temple with moderately dense, fine, more or less distinct pits; propleurum with dense, fine pits, slightly rugulose on lower part; mesopleurum with variable, fine pits and some rugulosity. *Shape*: Clypeus slightly upturned apically (Fig. 78:B); flagellum slender, narrowed apically, first 3 segments slightly elongate (Fig. 80:E,I); areola broadly hexagonal; area dentipara distinctly narrowed and elongate apically, apophysis distinct; 1st abdominal tergum moderately stout, postpetiole widely expanded, about twice as wide as long, sides parallel; femur III moderately swollen (Fig. 79:E), ratio about 0.28. *Other*: 3rd valvula moderately short, about as long as basal 3 hind tarsomeres; wings darkened.

REMARKS. This species appears most closely related to *subclavatus* and *patulus*. In addition to its western range, male *hesperus* is distinguished from males of the former by the black hind tarsomeres and the postpetiole and 2nd tergum with moderately dense, finely distinct pits. It differs from *patulus* by the elongate, finely punctate postpetiole. Female *hesperus* differs from females of both species by the orange flagellum, not bicolored, and by the dense, fine punctation of the frons, never coarse and rugulose. See additional discussion under *paludicola*.

RANGE. Pacific Northwest from northern California to British Columbia, eastward to Idaho and Montana. Collections May through August.

HOST. Diprionidae: *Neodiprion tsugae* Middleton, hemlock sawfly (OSU, USNM); *Neodiprion* sp. (USNM). Furniss and Dowden (1941: 49, 51) listed *subclavatus* as a cocoon parasite of the hemlock sawfly, but I have seen the reared specimens from Sweet Home, Oregon, and they are *hesperus*.

MATERIAL EXAMINED. **Holotype**: male, Mt. Rainier, 4000 ft., WASHINGTON, 14-viii-1940, H. and M. Townes (AEI). **Paratypes**: 173 M, 22 F (AEI, CAS, CNC, OSU, UCB, UCR): BRITISH COLUMBIA, CALIFORNIA, OREGON, WASHINGTON. Other material studied: 303 M, 14 F (AEI, CAS, CNC, Dasch, UCB, UCD, UCR, USNM): BRITISH COLUMBIA, CALIFORNIA, IDAHO, MONTANA, OREGON, WASHINGTON, WYOMING.

ETYMOLOGY. (Greek) *Hesperus* (western), referring to its ubiquity in the Pacific West.

Endasys inflatus (Provancher)
Renewed Status and New Combination
(Fig. 37:A-D; Map 9)

Ichneumon inflatus Provancher, 1875: 24, 83. **Lectotype** designated by Barron, 1975: 588-589.

Phygadeuon inflatus: Provancher, 1879: 75; 1882: 334, 336; 1883: 321, 773, 775; 1886: 47, 50; Davis, 1898: 348; Dalla Torre, 1902: 688; Nason, 1905: 150; Johnson, 1927: 142; Townes, 1944: 214-215; Carlson, 1979: 402.

Stylocryptus inflatus: Cushman, 1928: 928.

MALE DIAGNOSIS. Medium large, 6-7 mm long; face very finely granular with very dense, fine pits; clypeus yellow with margin sharp and more or less upturned (Fig. 78:C); coxae I and II and trochanters I and II mostly white, coxa III mostly orange, white apically and on trochanters; propleurum punctation moderately dense to moderately sparse with distinct pits; 1st abdominal tergum moderately slender (Fig. 84:F), postpetiole longer than wide; femur III orange with black apex, distinctly swollen (Fig. 79:D), ratio 0.25-0.28; propodeum with strong carinae, area dentipara distinctly narrowed, apophysis strong and vertically projecting; vertex slightly lengthened with occipital carina about the diameter of an ocellus distant from collar; eastern half of United States and southern Canada.

MALE DESCRIPTION. *White*: Tegula, coxae I and II except basally, coxa III apically, and trochanters. *Yellow*: Scape, clypeus, mandible, coxae I and II basally, femora I and II, tibiae I and II except latter dorsally pale yellow. *Orange*: Flagellum ventrally pale orange, coxa III except apically and sometimes blackish basally, femur III and tibia III except apically, hind basitarsus, abdomen except basally, apically, and often variably on 2nd tergum. *Black*: Flagellum dorsally, femur III and tibia III apically, most of hind tarsomeres except hind basitarsus, abdomen basally, apically, and variably on 2nd tergum. *Punctation* (Fig. 86:A-C,E): Face very finely granular with very dense, very fine pits; frons a little rugulose with dense pits; temple with dense, indistinct pits; propleurum with mostly moderately distinct pits, surface a little rugulose on lower part; mesopleurum impunctate centrally, otherwise with mostly dense, distinct pits, and rugulose; metapleurum distinctly rugulose; 1st and 2nd terga very slightly mat and very slightly wrinkled, otherwise mostly smooth and shiny. *Shape*: Clypeal margin sharp and more or less upturned (Fig. 78:C); flagellum moderately stout to moderately slender, its segments square or longer than wide on apical half; areola widely hexagonal, area dentipara distinctly narrowed, elongate apically (Fig. 83:C), apophysis strong and vertically projecting, propodeum rugulose between strong carinae; 1st tergum moderately stout to moderately slender, postpetiole squarish or longer than wide, 2nd tergum wide; femur III swollen (Fig. 79:D), ratio 0.25-0.28. *Other*: Vertex slightly lengthened (Fig. 89:B), occipital carina distant from collar by about diameter of an ocellus when viewed from side; wings hyaline; 5th and 6th abdominal terga membranous medially; glumes moderately dense.

FEMALE DIAGNOSIS. Medium large, 6-7 mm long; clypeus orange, margin sharp and distinctly upturned (Fig. 78:C); face finely granular with very dense, fine pits; frons rugulose with very dense, coarse pits; 1st abdominal tergum moderately slender (Fig. 84:C), postpetiole squarish or moderately widened, sides parallel; 3rd valvula moderately long, as long as basal 4 hind tarsomeres; femur III swollen (Fig. 79:E), ratio about 0.30; femur III and tibia III orange with black apices, trochanters yellow; vertex lengthened, occipital carina close to collar (Fig. 89:B).

FEMALE DESCRIPTION. *Yellow*: Flagellum basal half, flagellar segments 4-9 pale yellowish, scape, tegula, legs I and II, hind trochanters, sometimes coxa III ventrally. *Orange*: Clypeus except sometimes blackish orange, mandible, coxa III except sometimes ventrally, femur III and tibia III except often apically, and abdomen. *Black*: Clypeus sometimes more blackish orange, flagellum apical half, and often femur III and tibia III apically. *Punctation* (Fig. 86:E-F): Face finely granular with very dense, fine pits; frons rugulose with very dense, coarse pits; temple a little rugulose with dense, more or less coarse pits; propleurum with dense, finely distinct pits, surface finely rugulose; mesopleurum with variably dense to sparse, distinct pits, surface rugulose. *Shape*: Clypeal margin sharp and distinctly upturned; flagellum moderately stout, more or less linear to apex, basal 3 segments longer than wide (Fig. 80:F,H); areola broadly hexagonal, area dentipara distinctly narrowed, apophysis distinct and vertically projecting, propodeal carinae weak; 1st tergum moderately slender (Fig. 84:C), postpetiole mostly square or moderately widened to about 1.5 times as wide as long; femur III swollen (Fig. 79:E), ratio about 0.30; Rs straight. *Other*: 3rd valvula moderately long, as long as basal 4 hind tarsomeres; wings slightly tinted; vertex lengthened, occipital carina close to collar (Fig. 89:B).

REMARKS. This name has not been used since Cushman, 1928: 928. Davis, 1898: 348, incorrectly listed the name *Phygadeuon inflatus* as a synonym of *Zemiodes* (!) *seminiger* Harrington (now *Glyphicnemis californicus* (Cresson), Luhman, 1986: 133). This was probably because both species have a sharp and upturned clypeus, distinctly swollen hind femur, and often black on the apices of the hind femur and tibia. However, there are many important differences in structure since they are not congeneric.

Townes, 1944: 214, first lists *inflatus* among the synonyms of *subclavatus*. This was probably because of the similarity of the color pattern of the 2 species. However, *inflatus* clearly differs by the clypeal margin sharp and upturned, very fine rugulosity of the face, the coarser punctation of the frons, the strong carinae of the propodeum, and the 2nd abdominal tergum often blackish. Often the most conspicuous feature of the species is the one for which Provancher named it, the greatly inflated hind femur. The color pattern differs from *subclavatus* by having apices of both hind femur and tibia black, black hind tarsomeres, and sometimes with black on tergum 2. Darker forms of *inflatus* resemble *pubescens,* but the latter has mostly black hind coxa and femur, and the clypeal margin is not upturned.

The closest relative is Palearctic *testaceus* (Taschenberg). Males of this species differ by tergum 2 always orange, and the smaller size. Females differ from *inflatus* by the mostly orange flagellum, and the smaller size. See additional remarks under *tyloidiphorus*.

RANGE. Eastern half of the United States and southern Canada. Collections May through September.

HOST. Tenthredinidae: Nematinae: (prob.) *Nematus* sp., "locust sawfly" (USNM).

MATERIAL EXAMINED. **Lectotype**: male, QUEBEC, 1875, Provancher (UL). Paralectotype: female, QUEBEC, 1875, Provancher (UL): not congeneric. Other material studied: 299 M, 86 F (AEI, CAS, CNC, Dasch, UM, USNM): LOUISIANA, MARYLAND, MASSACHUSETTS, MICHIGAN, MINNESOTA, MISSOURI, MAINE, NEBRASKA, NEW BRUNSWICK, NEW JERSEY, NEW YORK, NORTH CAROLINA, OHIO, OKLAHOMA, ONTARIO, PENNSYLVANIA, QUEBEC, RHODE ISLAND, SOUTH CAROLINA, TENNESSEE, VIRGINIA, WASHINGTON D.C., WEST VIRGINIA.

ETYMOLOGY. *Inflatus* (swollen), probably referring to the swollen hind femur .

Endasys michiganensis Luhman, sp. n.
(Fig. 67:A-D; Map 10)

MALE DIAGNOSIS. Medium small, 5-5.5 mm long; face finely granular, surface shiny; scape yellow-orange, clypeus and mandible blackish orange, coxae black with trochanters yellow-orange, femora I and II and tibiae I and II yellow-orange, femur III mostly blackish, abdomen weakly bicolored, black basally and apically, and often some blackish on terga 4-6; flagellum stout to moderately stout, segments transverse or square on apical half; radial cell usually very slightly curved, about equal to length of 2nd-discoidal cell; femur III moderately swollen, ratio 0.20-0.22; eastern half of northern United States and southern Canada.

MALE DESCRIPTION. *Yellowish orange*: Flagellum ventrally, scape, sometimes clypeus more yellowish, tegula, trochanters, femora I and II, tibiae I and II except femora with some blackish, femur III basally, tibia III basally, sometimes mostly orange with some blackish laterally or apically; pale forms with foregoing areas yellowish. *Orange*: Clypeus and mandible more blackish orange, most of abdomen except basally and apically, often some blackish on terga 4-6. *Black*: Most of flagellum except ventrally, sometimes clypeus and mandible, coxae, femora I and II often variably blackish, most of femur III except basally, tibia III at least apically, hind tarsomeres except basitarsus more orangish, abdomen basally and apically, and often some blackish on terga 4-6; dark forms with blackish 1st-trochanters and most of abdomen except variably on tergum 2 apically and tergum 3 basally. *Punctation* (Fig. 86:B-C,E): Face finely granular with dense pits, surface shiny; frons with moderately sparse to moderately dense, finely distinct pits; temple with sparse, indistinct pits; propleurum sparsely punctate to impunctate, mostly smooth and shiny; mesopleurum impunctate centrally, otherwise with sparse, distinct pits, a little rugulose on lower part; 1st and 2nd terga very slightly mat or shiny. *Shape*: Clypeal margin sharp but not upturned (Fig. 78:B); flagellum stout to moderately stout (Fig. 80:D), its segments wider than long or squarish on apical half; areola broadly hexagonal, area dentipara narrowed, apophysis distinct, propodeal carinae moderately distinct; 1st tergum moderately slender (Fig. 84:F), postpetiole slightly longer than wide, 2nd tergum wide; femur III moderately swollen (Fig. 79:C), ratio 0.20-0.22; Rs very slightly curved, a little swollen. *Other*: Wings a little tinted; 5th abdominal sternum complete at least apically (Fig. 81:B); glumes dense.

FEMALE DIAGNOSIS. Medium small, 5-5.5 mm long; coxae blackish, trochanters yellow-orange, femur III black, tibia III orange with black apex, abdomen weakly bicolored, terga 3-apex blackish orange; flagellum moderately stout, more or less linear to apex, basal 3 segments short (Fig. 80:F,H); face and clypeus black and shiny with densely distinct pits; Rs more or less curved, radial cell short (Fig. 82:D).

FEMALE DESCRIPTION. *Orange*: Flagellum basal half, scape, mandible, tegula, trochanters, femora I and II and tibiae I and II except latter apically, abdomen on 1st tergum apically, 2nd entirely, and terga 3-apex orangish black. *Black*: Flagellum basal half, clypeus, coxae I and II more orangish black, coxa III, femur III, tibia III apically, and abdomen basally and blackish on terga 3-apex. *Punctation* (Fig. 86:B-C,E): Face shiny with dense, fine pits; frons with mostly dense, distinct pits; temple with sparse, indistinct pits; propleurum with dense, fine pits; mesopleurum with mostly sparse, finely distinct pits, surface slightly rugulose. *Shape*: Clypeal margin sharp but not upturned (Fig. 78:B); flagellum moderately stout, more or less linear to apex, basal 3 segments short, nearly moniliform (Fig. 80:F,H); areola broadly hexagonal, area dentipara distinctly narrowed, apophysis distinct, propodeal carinae distinct; 1st tergum moderately stout (Fig. 84:B), postpetiole widened to about 1.8 times its length; femur III moderately swollen (Fig. 79:D), ratio 0.28; radial cell short, Rs more or less curved (Fig. 82:D). *Other*: 3rd valvula moderately long, about as long as basal 4 hind tarsomeres; wings tinted.

REMARKS. See discussion under *brevicornis*.

RANGE. Minnesota, Michigan, and Ontario eastward to Newfoundland, New Brunswick, Maine, and New Jersey. Collection May through August.

MATERIAL EXAMINED. **Holotype**: male, Yellow Dog Plains, Marquette Co., MICHIGAN. 28-vi-1961, H. and M. Townes (AEI). **Paratypes**: 54 M, 12 F (AEI, CNC, Dasch): MAINE, MASSACHUSETTS, MICHIGAN, MINNESOTA, NEW BRUNSWICK, NEWFOUNDLAND, NEW JERSEY. Other material studied: 94 M (AEI, CNC): ALBERTA, MICHIGAN, NEW BRUNSWICK, NEWFOUNDLAND, ONTARIO, QUEBEC.

ETYMOLOGY. From Michigan, the type locality.

Endasys nemati Luhman, sp. n.
(Fig. 65:A-D; Map 14)

MALE DIAGNOSIS. Medium size, 4.5-6.5 mm long; coxae black basally, white apically, trochanters white, abdomen black basally and apically, and variably on terga 2-6, remaining areas orangish, femur III and tibia III mostly black, often more yellowish basally; face finely rugulose or finely granular with dense pits; clypeus yellow and flat, margin sharp with weakly upturned edge (Fig. 78:B); propodeal carinae moderately distinct, area dentipara narrowed, apophysis moderately distinct; 1st tergum moderately slender or moderately stout, postpetiole longer than wide or squarish; femur III moderately swollen (Fig. 79:C), ratio 0.22-0.24; wings more or less hyaline; western North America.

MALE DESCRIPTION. *White*: Tegula sometimes more whitish, coxae apically, trochanters. *Yellow*: Scape, clypeus, mandible, usually tegula, femora I and II, tibiae I and II, femur III often more yellowish basally, tiba III basal half. *Orange*: 1st tergum apically, variably on 2nd, and at least basally on terga 3-6; larger forms with mostly orange abdomen except basally and apically. *Black*: Flagellum, coxae except apically, most of femur III except often basally, tibia III at least apical half, hind tarsomeres, and abdomen at least basally and apically, often variably on tergum 2, and apically or laterally on terga 3-6; small forms with abdomen entirely black. *Punctation* (Fig. 86:B,E): Face finely granular or finely rugulose with dense, fine pits; frons with moderately dense to moderately sparse, dense, fine, more or less distinct pits; temple with moderately sparse, fine, indistinct pits; propleurum with dense to sparse, distinct pits; mesopleurum impunctate centrally, otherwise with sparse, distinct pits, surface a little rugulose on lower part; 1st and 2nd terga mat and very slightly wrinkled. *Shape*: Clypeus flat, margin sharp but not distinctly upturned (Fig. 78:B); flagellum moderately stout (Fig. 80:C), its segments squarish on apical half; areola broadly hexagonal, area dentipara narrowed, apophysis moderately strong, propodeal carinae moderately distinct; 1st tergum moderately slender or moderately stout, postpetiole longer than wide or square, 2nd tergum slightly widened; femur III moderately swollen (Fig. 79:C), ratio 0.22-0.24; Rs very slightly curved. *Other*: Sometimes 3rd tyloid on 9th flagellar segment; wings more or less hyaline; 5th abdominal sternum membranous medially; glumes sparse.

FEMALE DIAGNOSIS. Small, 4-5 mm long; abdomen black basally and apically, and variably on terga 2-6, usually laterally and apically, remaining areas orangish, legs yellowish orange except coxa III and femur III more blackish orange, and tibia III often blackish apically; flagellum orangish, short and stout, mostly linear to apex, basal 3 segments nearly moniliform (Fig. 80:F,G); face finely rugulose with very dense pits, frons with variably sparse to dense, finely distinct pits; hind femur swollen (Fig. 79:E), ratio 0.28-0.29; 3rd valvula moderately long, as long as basal 4 hind tarsomeres; wings more or less hyaline, Rs slightly curved (Fig. 82:D); clypeus orange, impressed apically, margin not distinctly upturned; area dentipara narrowed with moderately strong apophysis.

FEMALE DESCRIPTION. *Yellowish orange*: Flagellum, scape, clypeus, mandible, tegula, legs except coxa III, femur III, and tibia III apically, 1st abdominal tergum apically, variably on terga 2-6 except laterally and apically; larger forms with abdomen mostly orange except basally and apically. *Black*: Coxa III except apically more orangish black, femur III orangish black, tibia III apically more blackish, abdomen at least basally and apically, variably on terga 2-6 laterally and apically; small forms with abdomen mostly black. *Punctation* (Fig. 86:B): Face finely rugulose with very dense pits; frons with variably sparse to dense, finely distinct pits; temple with moderately sparse, more or less fine pits; propleurum with moderately dense, finely distinct pits; mesopleurum with very sparse, very fine pits, surface a little rugulose. *Shape*: Clypeus flat, margin weakly upturned (Fig. 78:B); flagellum short and stout, segments linear to apex, basal 3 segments nearly moniliform (Fig. 80:F-G); areola broadly hexagonal, area dentipara narrowed, apophysis distinct, propodeum more or less level dorsally, carinae weak; 1st tergum moderately stout (Fig. 84:B), postpetiole gradually widened, about 1.8 times as wide as long; femur III swollen (Fig. 79:E), ratio 0.28-0.29; radial cell short, Rs often slightly curved, at least apically. *Other*: 3rd valvula moderately long, as long as basal 4 hind tarsomeres; wings more or less hyaline.

REMARKS. This species may be related to eastern *brevicornis*. Males are distinguished by the mostly black hind leg and 1st and 2nd abdominal terga. Females are separate by the orangish flagellum, basal 3 segments moniliform, and the yellowish orange legs I and II.

RANGE. Alaska to northern California, eastward to Wyoming, Idaho, and Minnesota. Collections May through September.

HOST. Tenthredinidae: Nematinae: *Nematus currani* Ross (= *nigriventris* Curran), a sawfly of *Populus* (CNC).

MATERIAL EXAMINED. **Holotype**: male, Taft, BRITISH COLUMBIA, 29-v-1944, C.V.G. Morgan (CNC). **Paratypes**: 36 M, 16 F (AEI, Dasch, CNC): ALASKA, ALBERTA, BRITISH COLUMBIA, CALIFORNIA, IDAHO, OREGON, WASHINGTON, WYOMING, YUKON TERRITORY. Other material studied: 30 M, 11 F (AEI, CNC, UM): BRITISH COLUMBIA, CALIFORNIA, MINNESOTA, OREGON, UTAH.

ETYMOLOGY. From *Nematus* the tenthredinid host genus from which the type and much of the type series were reared.

Endasys paludicola (Brues)
(Fig. 20:A-D; Map 6)

Oxytorus paludicola Brues, 1908: 50. **Lectotype**, here designated.
Endasys (Endasys) paludicola: Townes, 1944: 213; Townes and Townes, 1951: 246.
Endasys paludicola: Carlson, 1979: 417.

MALE DIAGNOSIS. Medium large, 6-7 mm long; coxae mostly blackish, often yellowish apically; yellowish scape, clypeus (often more blackish), and trochanters; femur III and III black at least apically; abdomen black basally and apically, and variably orangish or blackish on terga 2-6; 1st and 2nd terga smooth and shiny, 1st tergum moderately slender (Fig. 84:F), postpetiole longer than wide or square; propodeal carinae moderately strong, area dentipara regular, apophysis distinct; femur III moderately swollen to moderately slender, ratio 0.21-0.23; frons evenly punctate with dense, very fine pits; western United States

MALE DESCRIPTION. *Yellow*: Scape, clypeus except more blackish yellow, coxae I and II apically, trochanters, femora I and II, tibiae I and II. *Orange*: Coxa III except basally, femur III, tibia III except at least apically, variably on terga 1-6. *Black*: Coxae basally, femur III and tibia III at least apically, hind tarsomeres, abdomen apically and variably on terga 1-6. *Punctation* (Fig. 86:B-C): Face finely rugulose with very dense, fine pits; frons with evenly dense, finely distinct pits; temple with indistinct pits, hair sockets dense; propleurum with dense to sparse, finely distinct pits; mesopleurum impunctate centrally, otherwise with very sparse, finely dense pits, surface smooth and shiny; 1st and 2nd terga smooth and shiny. *Shape*: Clypeal margin not upturned apically (Fig. 78:A); flagellum moderately slender (Fig. 80:B), its segments longer than wide to apex; areola broadly hexagonal, area dentipara regular or slightly narrowed, apophysis distinct, propodeal carinae moderately strong; 1st tergum moderately slender(Fig. 84:F), postpetiole longer than wide, 2nd tergum regular; femur III moderately slender to moderately swollen, ratio 0.21-0.23; Rs mostly straight, slightly curved at apex. *Other*: Wings tinted, 5th abdominal sternum more or less complete apically, and glumes moderately dense (Fig. 88:B).

FEMALE DIAGNOSIS. Medium, 5-6 mm long; coxae II and III blackish basally, femur III orange with blackish, tibia III and hind tarsomeres orange, legs I and II yellowish orange; abdomen weakly bicolored, blackish basally, 2nd tergum orange except laterally blackish, 3rd tergum more brownish orange, and blackish on terga 4 to apex of abdomen; scape yellowish orange, flagellum mostly yellowish orange, paler on basal half, slender, narrowed apically, basal 3 segments short (Fig. 80:E,H); clypeus blackish with weakly upturned margin; punctation generally moderately dense to moderately sparse with fine pits, surfaces smooth and shiny; apophysis distinct; 1st abdominal tergum moderately stout (Fig. 84:B); 3rd valvula moderately short, as long as basal 3 hind tarsomeres.

FEMALE DESCRIPTION. *Yellowish orange*: Scape, mandible, flagellum (paler basal half), tegula, legs I and II except coxa II basally, coxa III ventroapically, and hind trochanters. *Orange*: Femur III mostly, tibia III and tarsomeres, 1st tergum apically, 2nd tergum except laterally, and 3rd tergum more brownish orange. *Black*: Clypeus except apically orangish black, coxa III basally, 1st tergum except apically, 2nd tergum laterally, and terga 4 to apex of abdomen. *Punctation* (Fig. 86:B-D): Face shiny with moderately dense, variably distinct pits; frons with moderately sparse, fine pits; temple with moderately sparse, indistinct pits; propleurum with moderately dense, very fine pits; mesopleurum with moderately sparse, variable pits. *Shape*: Clypeal margin weakly upturned (Fig. 78:B); flagellum slender, narrowed apically, basal 3 segments short (Fig. 80:E,H); areola broadly hexagonal, area dentipara slightly narrowed, a little elongate apico-laterally (Fig. 83:C), apophysis distinct, propodeal carinae distinct; 1st abdominal tergum moderately stout (Fig. 84:B), postpetiole about 1.5 times as wide as long; femur III moderately swollen (Fig. 79:D), ratio 0.25; Rs nearly straight. *Other*: 3rd valvula moderately short, as long as basal 3 hind tarsomeres; wings slightly tinted.

REMARKS. This species appears most closely related to *albior*. Males are distinguished by the hind femur and tibia mostly orange with black apically, hind femur more swollen (ratio 0.21-0.23). It may have some affinity to *hesperus* from which males differ by having the hind coxa partly black, the postpetiole and 2nd tergum without finely distinct pits, and usually terga 2-6 partly black. Females differ from both species by the bicolored flagellum, basal 3 segments short. Additionally, females differ from *albior* by the orangish hind femur and shorter 3rd valvula (as long as basal 3 hind tarsomeres); and from *hesperus* by the partly blackish abdomen and yellowish front and middle legs, not orange.

RANGE. Idaho, Montana, Wyoming, and Colorado. Collections July and August.

MATERIAL EXAMINED. **Lectotype**: male, Florissant, COLORADO, 26-vii-1908, Cockerell (MPM). **Paralectotype**: (same as lectotype; MCZ). Other material studied: 23 M, 1 F (AEI, Dasch, FLDA, USNM): COLORADO, IDAHO, MONTANA, WYOMING.

ETYMOLOGY. *Paludi-* (swamp) + *cola* (dweller) referring to the habitat of the type locality.

Endasys patulus (Viereck)
Renewed Status and New Combination
(Fig. 66:A-D; Map 21)

Phygadeuon (Bathymetis) patulus Viereck, 1911: 193. **Holotype** by monotypy, here determined. Swenk, 1911: 29; Townes, 1944: 215; Townes and Townes, 1951: 246; Carlson, 1979: 418.

MALE DIAGNOSIS. Large, 7-8.5 mm long; flagellum long and slender, tyloids long (Fig. 87:A); femur III slender (Fig. 79:B), ratio about 0.20, orange with black apex; petiole long and slender, postpetiole squarish and flattened, sides nearly parallel; area dentipara narrowed or regular, widened laterally, apophysis moderately distinct, propodeal carinae distinct; clypeal margin very slightly upturned medially, sometimes giving clypeus slightly beaked appearance; frons with dense to very dense, finely distinct pits; coxa I mostly white, coxa II orange with white apex, coxa III entirely orange, trochanters white, tibia III always black apically, hind tarsomeres always black, abdomen orange, sometimes black basally and on parameres; eastern half of United States and southern Canada.

MALE DESCRIPTION. *White*: Scape, usually clypeus and mandible, tegula, coxa I except basally, coxa II apically, trochanters, and tibiae I and II dorsally. *Yellow*: Often clypeus and mandible, coxa I basally, tibiae I and II except dorsally. *Orange*: Middle coxa except apically, femora I and II, tibiae I and II except dorsally; orange forms with mostly orange thorax and propodeum. *Black*: Flagellum except more orangish black ventrally, femur III and tibia III apically, hind tarsomeres, usually abdomen basally, sometimes parameres. *Punctation* (Fig. 86:B,D): Face finely rugulose or finely granular with very dense, fine pits; frons with dense to very dense, finely distinct pits, surface sometimes slightly rugulose; temple with dense, fine, more or less indistinct pits; propleurum with dense to sparse, fine pits; mesopleurum mostly impunctate, otherwise with moderately sparse to moderately dense punctation, finely distinct pits, surface slightly rugulose; 1st and 2nd terga smooth and shiny. *Shape*: Clypeal margin slightly upturned medially, often giving beak-like appearance; flagellum elongate and slender (Fig. 80:A), its segments distinctly elongate to apex, apical segment long and slender; areola broadly to widely hexagonal, area dentipara narrowed or regular, widened laterally, apophysis more or less distinct, propodeal carinae distinct, sometimes weak; petiole long and slender, postpetiole squarish (Fig. 84:F), sides nearly parallel, 2nd tergum regular or wide; femur III slender (Fig. 79:B), ratio 0.20. *Other*: Wings tinted, 5th and 6th abdominal terga membranous medially, glumes dense, tyloids (2) elongate (Fig. 87:A).

FEMALE DIAGNOSIS. Large, 6-8 mm long; 3rd valvula short, about as long as basal 2 hind tarsomeres; flagellum bicolored, slightly narrowed apically (Fig. 80:E), basal 3 segments elongate; clypeal margin slightly upturned medially, often giving slightly beak-like appearance; 1st abdominal tergum stout (Fig. 84:A), postpetiole distinctly widened, twice as wide as long, sides nearly parallel; areola widely to broadly hexagonal, area dentipara distinctly narrowed, elongate apico-laterally (Fig. 83:C), apophysis more or less distinct; femur III moderately swollen (Fig. 79:C), ratio 0.25-0.26.

FEMALE DESCRIPTION. *Orange*: Flagellum on basal half, scape, mandible, tegula, legs except often tibia III apically and hind tarsomeres, abdomen; orange forms with mostly orange thorax and propodeum. *Black*: Flagellum on apical half, clypeus, often tibia III apically, hind tarsomeres. *Punctation* (Fig. 86:E-F): Face, frons, propleurum, and mesopleurum rugulose with very dense, coarse pits; temple with dense to moderately dense, distinct pits. *Shape*: Flagellum long and moderately slender, slightly narrowed apically, basal 3 segments elongate (Fig. 80:E,I); clypeal margin slightly upturned medially, often giving clypeus slightly beaked appearance; areola broadly to widely hexagonal, area dentipara distinctly narrowed, elongate apico-laterally (Fig. 83:C), apophysis more or less distinct, propodeum level dorsally with weak carinae; 1st tergum stout and wide, postpetiole distinctly widened, twice as wide as long, sides parallel; femur III moderately swollen (Fig. 79:C), ratio 0.25-0.26; Rs straight. *Other*: 3rd valvula short, about as long as basal 2 hind tarsomeres; wings tinted.

REMARKS. This name has been raised from synonymy with *subclavatus*. It is superficially similar to the latter, but has a generally stouter body with more slender appendages. Males are distinguished by the black hind tarsomeres, long, slender flagellum, and the nearly square postpetiole with reduced dorsomedian carina. Females differ by the very short 3rd valvula, usually shorter than the basal 2 hind tarsomeres, and the widely expanded postpetiole, twice as wide as long. For affinities of *patulus*, see remarks under *hesperus*, *arkansensis* and *aurarius*.

RANGE. Eastern half of the United States and southern Canada.

HOST. Diprionidae: *Zadiprion townsendi* (Cockerell) (Swenk, 1911: 29) (USNM); *Neodiprion excitans* Rohwer, blackheaded pine sawfly (USNM); *N. lecontei* (Fitch), redheaded pine sawfly (CNC); *N. merkeli* Ross, slash pine sawfly (FLDA, USNM); *N. nannulus* Schedle, red pine sawfly (CNC); *N. pratti banksianae* Rohwer, jack pine sawfly (CNC); *N. pratti* (Dyar), Virginia pine sawfly (FLDA); *N. sertifer* (Geoffroy), European pine sawfly (CNC); *N. swainei* Middleton, Swain jack pine sawfly (CNC); *N. taedae linearis* Ross, loblolly pine sawfly (UALM, UCR). Tenthredinidae: Nematinae: *Pristiphora erichsonii* (Hartig), larch sawfly (CNC).

Most specimens of *Endasys* reared from the above hosts were labelled as *subclavatus* in the CNC, so it is possible that virtually all host records listed with it in the literature belong with *patulus*. No reared material in any collection examined contained *subclavatus*. Therefore, the following hosts (ascribed to *subclavatus*) are listed from the literature, in addition to the above: Diprionidae: *Gilpinia hercyniae* (Hartig), European spruce sawfly (Reeks, 1938: 27; Peirson and Nash, 1940: 16-17; Brown, 1941: 5; Finlayson, 1963: 491); *Neodiprion* sp. (Peirson and Nash, 1940: 16-17); *N. abbottii* (Leach) (Hetrick, 1941: 376; Finlayson, 1963: 476, 491); *N. excitans* Rohwer (Walkley, 1967: 94), blackheaded pine sawfly; *N. lecontei* (Fitch), redheaded pine sawfly (Bentley, 1940; Benjamin, 1955; Finlayson, 1963: 479, 491); *N. nannulus nannulus* Schedl, red pine sawfly (Coppel, 1954: 168; Forbes et al., 1960; Underwood, 1960; Finlayson, 1963: 481, 491; Bobb, 1963: 618; Walkley, 1967: 94); *N. pratti banksianae* Rohwer, blackheaded jack pine sawfly (Griffiths, 1959: 508, 510; 1960: 656; Finlayson, 1963: 483, 491; Price, 1971: 590); *N. pratti pratti* (Dyar), Virginia pine sawfly (Bobb, 1963: 618-620); *N. sertifer* (Geoffroy), European pine sawfly (Girth and McCoy, 1946; Craighead, 1950; Finlayson and Finlayson, 1958; Griffiths, 1959: 508, 510; Finlayson, 1963: 491; Walkley, 1967: 94); *N. swainei* Middleton, Swaine jack pine sawfly (Tripp, 1960, 1961: 53; Finlayson, 1963: 484, 491; Walkley, 1967: 94); Price, 1971: 590; *N. taedae linearis* Ross (Walkley, 1958: 45; Finlayson, 1963: 491), loblolly pine sawfly; *N. virginianae* Rohwer, redheaded jack pine sawfly (Schedl, 1939; Finlayson, 1963: 485, 491).

MATERIAL EXAMINED. **Holotype:** male, Crawford, NEBRASKA, emerged 24-vii-1910, M. Swenk (USNM). No paratypes. Other material studied: 339 M, 224 F (AEI, CNC, Dasch, FLDA, MCZ, UCB, UCD, UCR, UM, USNM): ALABAMA, ARKANSAS, CONNECTICUT, FLORIDA, ILLINOIS, INDIANA, LOUISIANA, MAINE, MARYLAND, MASSACHUSETTS, MICHIGAN, MINNESOTA, NEBRASKA, NEWFOUNDLAND, NEW HAMPSHIRE, NEW JERSEY, NEW YORK, NORTH CAROLINA, NOVA SCOTIA, OHIO, ONTARIO, PENNSYLVANIA, QUEBEC, SOUTH CAROLINA, SOUTH DAKOTA, TENNESSEE, VIRGINIA, WASHINGTON D.C., WEST VIRGINIA, WISCONSIN.

ETYMOLOGY. *Patulus* (spreading wide, from *patere* "to spread"), probably referring to the very wide postpetiole.

Endasys praerotundiceps Luhman, sp. n.
(Fig. 63:A-D; Map 17)

MALE DIAGNOSIS. Small, 3.5-5.5 mm long; abdomen mostly black, sometimes terga 3-6 yellowish basally; coxae I and II mostly white, trochanters white, coxa III yellow or black with white ventro-apically, femur III and tibia III black; tibia III with 3 rows of short, white spines; 1st abdominal tergum moderately stout (Fig. 84:E), postpetiole squarish, 2nd tergum wide; femur III moderately swollen (Fig. 79:C), ratio about 0.24; eastern half of United States and southern Canada.

MALE DESCRIPTION. *White*: Scape, clypeus, mandible, tegula, most of coxae I and II except basally, coxa III ventro-apically, trochanters. *Yellow*: Flagellum ventrally more yellowish; femora I and II, tibiae I and II, coxae I and II basally; coxa III variably, sometimes entirely yellow except ventro-apically; femur III more or less basally; sometimes basally on terga 3-6. *Black*: Flagellum at least dorsally; coxa III variably except variable yellow and white ventro-apically, sometimes entirely black except ventro-apically; femur III except often yellowish basally; tibia III, hind tarsomeres; most of abdomen except often yellow basally on terga 3-6. *Punctation* (Fig. 86:B-C,E): Face finely rugulose with very dense pits; frons with moderately dense, distinct pits; temple with moderately dense, indistinct pits; propleurum densely punctate to impunctate, indistinct pits; mesopleurum impunctate centrally and a little rugulose, otherwise with mostly sparse, indistinct pits, surface a little rugulose; 1st and 2nd terga mostly smooth and shiny, pits sparse and very fine. *Shape*: Clypeal margin sharp but only faintly upturned (Fig. 78:B); flagellum moderately stout (Fig. 80:C), its segments a little longer than wide; areola widely hexagonal, area dentipara nearly regular or slightly narrowed, apophysis distinct, propodeal carinae distinct; 1st tergum moderately stout (Fig. 84:E), postpetiole squarish, 2nd tergum wide; femur III moderately swollen (Fig. 79:C), ratio 0.24; Rs very slightly curved. *Other*: Wings more or less hyaline; 5th abdominal sternum membranous medially; hind tibial spines short, white, and in 3 rows; glumes dense (Fig. 88:A).

FEMALE DIAGNOSIS. Small, 4-5 mm long; 1st abdominal tergum blackish basally with yellowish apically, remaining terga variably blackish yellow, black at least laterally; hind trochanters and coxa yellow, latter at least ventro-apically, femur III blackish with yellow basally and apically; flagellum bicolored yellow and black, moderately stout, linear to apex (Fig. 80:F); Rs a little curved, radial cell short (Fig. 82:D); femur III moderately swollen (Fig. 79:D), ratio about 0.28; 3rd valvula moderately long, about as long as basal 4 hind tarsomeres; propodeum more or less level dorsally, area dentipara narrowed, apophysis distinct; 1st tergum moderately slender (Fig. 84:C), postpetiole gradually widened, about twice as wide as long; punctation of frons moderately dense and distinct, temple with moderately sparse, more or less indistinct pits; clypeus black or yellowish, margin sharp and very slightly upturned (Fig. 78:B).

FEMALE DESCRIPTION. *Yellow*: Flagellum basal half, scape often, mandible, tegula, coxae except coxa III often more blackish yellow, trochanters, legs I and II, femur III basally and apically, tibia III basally, hind tarsomeres more blackish yellow, 1st tergum apically, and variably blackish yellow on remaining terga, sometimes mostly yellow except laterally. *Black*: Flagellum basal half, usually clypeus blackish, femur III yellowish black except basally and apically, tibia III yellowish black except basally, 1st tergum except apically, and variably blackish yellow on terga 2-apex, at least laterally, sometimes mostly yellowish black. *Punctation* (Fig. 86:B-E): Face rugulose with dense pits; frons with moderately dense, distinct pits; temple with moderately sparse or moderately dense punctation, more or less indistinct pits; propleurum slightly rugulose with moderately dense to moderately sparse, fine pits; mesopleurum with mostly sparse, distinct pits, surface rugulose. *Shape*: Clypeal margin sharp and very slightly upturned (Fig. 78:B); flagellum moderately stout, linear to apex, basal 3 segments a little longer than wide (Fig. 80:F,H); areola broadly hexagonal, area dentipara a little narrowed, apophysis distinct, propodeum more or less level dorsally, carinae weak; 1st tergum moderately slender (Fig. 84:C), postpetiole gradually widened, about 1.5 times as wide as long; femur III moderately swollen (Fig. 79:D), ratio about 0.28; radial cell short, Rs a little curved (Fig. 82:D). *Other*: 3rd valvula moderately long, about as long as basal 4 hind tarsomeres; wings tinted.

REMARKS. This species has been called *rotundiceps* Provancher, and may well have been the species he intended to describe. However, the specimen chosen by Barron (1975: 445-446) as lectotype is not the same species as the 3 other specimens included with it in Provancher's collection (UL). Therefore, *praerotundiceps* is described here to embrace the species originally seen in these 3 specimens. Fortunately, little material in collections was labelled *rotundiceps*, so there will be little confusion. The new species is much more commonly collected than the lectotype species. See additional remarks under *rotundiceps* and *subclavatus*. For affinities of *praerotundiceps*, see remarks under *pubescens*.

RANGE. Eastern half of the United States and southern Canada, and Colorado. Collections May through August.

HOST. Tenthredinidae: Nematinae: *Pristiphora geniculata* (Hartig), mountain ash sawfly (CNC); Heterarthrinae: *Caliroa cerasi* (Linnaeus), pear slug on cherry leaves (USNM); *Metallus rohweri* MacGillivray, a leafminer of *Rubus* (USNM); *Profenusa* sp. on white birch, a birch leafminer (USNM); probably also *Profenusa* on a red oak (USNM). The latter 2 hosts were reared from soil cages under Profenusa-infested white birch and red oak.

MATERIAL EXAMINED. **Holotype**: male, Chaffeys Locks, ONTARIO, 9-vii-1975, J. Belwood (AEI). **Paratypes**: 118 M, 78 F (AEI, CNC, Dasch, USNM): CONNECTICUT, IOWA, KANSAS, MAINE, MARYLAND, MASSACHUSETTS, MICHIGAN, NEW BRUNSWICK, NEW JERSEY, NEW YORK, NORTH CAROLINA, NOVA SCOTIA, OHIO, ONTARIO, PENNSYLVANIA, QUEBEC, RHODE ISLAND, SOUTH CAROLINA, VERMONT, VIRGINIA WEST VIRGINIA, WISCONSIN. Other material studied: 640 M, 104 F (AEI, Dasch, OSU, UCB, UM, USNM): CONNECTICUT, COLORADO, KANSAS, MAINE, MARYLAND, MICHIGAN, MINNESOTA, NEW BRUNSWICK, NEW HAMPSHIRE, NEW JERSEY, NEW YORK, NORTH CAROLINA, OHIO, ONTARIO, PENNSYLVANIA, QUEBEC, RHODE ISLAND, SOUTH CAROLINA, TENNESSEE, VERMONT, VIRGINIA, WEST VIRGINIA, WISCONSIN.

ETYMOLOGY. *Prae-* (very, true) + *rotundiceps* (species name), referring to its historical association with the name *rotundiceps* Provancher.

Endasys pubescens (Provancher)
(Fig. 55:A-D; Map 16)

Cryptus pumilus Provancher, 1874: 177, 283 (misidentified by Provancher, not congeneric).

Phygadeuon pubescens Provancher, 1874: 282; 1875: 183; 1879: 72; 1882: 335; 1883: 318, 775; 1886: 50; 1887: 34. **Holotype** (by monotypy) cited by Gahan and Rohwer (1918: 166) and Barron (1975: 537); Townes, 1944: 213-214; Townes and Townes, 1951: 246. Lists: Riley and Howard, 1890: 153; Dalla Torre, 1902: 692; Johnson, 1927: 142; Brimley, 1942: 30.

Phygadeuon dubius Provancher, 1874: 283, 1875: 183; Dalla Torre, 1902: 692; Barron, 1975: 468 (type not found).

Alomya pulchra Provancher, 1875: 120. (Holotype, cited by Barron, 1975: 537.)

Phygadeuon pallicoxus Provancher. 1879: 75; 1882: 336, 1883: 775, 1886: 50; (Lectotype designated by Gahan and Rohwer, 1918: 166, cited by Barron, 1975: 525); Townes, 1944: 213. List: Brimley, 1942: 30; Wray, 1967: 105.

Phygadeuon pallidicoxus Dalla Torre, 1902: 690 (emendation of *P. pallicoxus* Provancher).

Stylocryptus pubescens: Cushman (1926) 1928: 928. List: Johnson. 1927: 142.

Endasys (Endasys) pubescens: Townes, 1944: 213; Townes and Townes, 1951: 246.

Endasys pubescens: Carlson, 1979: 417. Biology: Houseweart and Kulman, 1976: 863-864; Rau, 1976: 25-26, 33, 35-36; Thompson and Kulman, 1980: 25-29.

MALE DIAGNOSIS. Medium large, 6-7 mm long; coxae I and II white, coxa III black with white apically and often ventrally, trochanters white, scape and clypeus white or yellow, femur III mostly black, tibia III mostly orange with black apically, abdomen generally orange with black basally, apically, and often variably on tergum 2; clypeal margin not upturned (Fig. 78:A); propodeum rugulose between strong, often coarse carinae, area dentipara distinctly narrowed, apophysis strong; femur III moderately swollen (Fig. 79:C), ratio 0.23-0.25; 1st tergum moderately stout (Fig. 84:E), postpetiole squarish, 1st and 2nd terga slightly mat and slightly wrinkled; wings more or less hyaline; Alaska and Yukon Territory to eastern United States and southern Canada.

MALE DESCRIPTION. *White*: Sometimes scape, clypeus, and mandible; tegula; coxae I and II at least apically, sometimes almost entirely; hind coxa apically, sometimes also ventrally; trochanters. *Yellow*: Usually scape, clypeus, and mandible; femora I and II except sometimes femur II blackish; tibiae I and II except pale yellow dorsally. *Orange*: Usually flagellum pale orange ventrally, often femur III at base, tibia III except at least apically, basally on hind basitarsus, abdomen except basally, apically, and variably on tergum 2. *Black*: Flagellum at least dorsally, most of femur III, tibia III at least apically, tibia III except basitarsus basally, abdomen basally and apically, and often variably on tergum 2; dark forms with terga 1 and 2 mostly black, and blackish on terga 3-6 apically and laterally. *Punctation* (Fig. 86:C-F): Face finely granular with very dense, fine pits; frons a little rugulose with dense to very dense pits; temple with moderately dense, indistinct pits; propleurum with dense to very sparse, variably distinct pits; mesopleurum impunctate centrally with rugulosity, otherwise with variably dense to sparse punctation, surface with rugulosity; postpetiole and 2nd terga slightly mat and slightly wrinkled. *Shape*: Clypeal margin not upturned (Fig. 78:A); flagellum moderately stout (Fig. 80:C), its segments squarish on apical half, only a little longer than wide on basal half; areola widely hexagonal, area dentipara distinctly narrowed, often elongate apically (Fig. 83:C), apophysis strong, propodeum rugulose between strong carinae; 1st tergum moderately stout (Fig. 84:E), postpetiole squarish, 2nd tergum wide; femur III moderately swollen (Fig. 79:C), ratio usually 0.23-0.25; Rs straight. *Other*: Wings more or less hyaline or slightly tinted; 5th abdominal sternum usually membranous medially, sometimes complete apically; glumes dense; tyloids (2) short (Fig. 87:D) and prominent.

FEMALE DIAGNOSIS. Medium large, usually 6-7 mm long; yellow coxae and trochanters, coxae more blackish basally, femora I and II yellow with black, tibia I and II pale yellow dorsally, femur III black, tibia III orange with black apex, abdomen mostly orange with black basally, apically, and laterally; clypeus black without upturned margin (Fig. 78:A); 3rd valvula moderately long, about as long as hind tarsomeres; flagellum bicolored, linear to apex (Fig. 80:F); punctation of head dense to very dense, face finely granular or finely rugulose, frons slightly rugulose, temple smooth with indistinct pits, propleurum with very dense punctate, fine pits; propodeal carinae distinct, area dentipara distinctly narrowed and elongate apicolaterally (Fig. 83:C), apophysis distinct.

FEMALE DESCRIPTION. *Yellow*: Scape, mandible, tegula, coxae I and II except usually basally, trochanters, coxa III ventrally, femur I and II usually with some blackish, tibiae I and II except pale yellow dorsally. *Orange*: Flagellum pale orange basal half, segments 5-9 more yellowish; clypeus sometimes orangish; tibia III except apically; hind tarsomeres except usually apices black; abdomen except basally, apically, and laterally. *Black*: Flagellum apical half, clypeus usually, coxae basally, femur III, tibia III apically, apices of hind tarsal segments, abdomen basally, apically, and laterally. *Punctation* (Fig. 86:B,D-E): Face finely rugulose or finely granular with very dense pits; frons a little rugulose with very dense pits; temple with dense, fine, more or less distinct pits; propleurum finely rugulose with very dense, fine pits; mesopleurum with variably dense to sparse punctation, fine, more or less distinct surface slightly rugulose. *Shape*: Clypeal margin not upturned (Fig. 78:A); flagellum stout, more or less linear to apex, basal 3 segments short (Fig. 80:F,H); areola broadly hexagonal, area dentipara distinctly narrowed and elongate, apophysis distinct, propodeal carinae distinct; 1st tergum moderately stout (Fig. 84:B), postpetiole gradually widened, about twice as wide as long; femur III swollen (Fig. 79:E), ratio 0.30; Rs straight (Fig. 82:B). *Other*: 3rd valvula long, about as long as hind tarsomeres; wings tinted.

REMARKS. Its closest relative appears to be western Palearctic *euxestus* (Speiser), which differs from *pubescens* by the distinctly elongate radial cell. In the Nearctic it appears related to *praerotundiceps*. It differs from it by the mostly orange abdomen and the larger size, over 5 mm long. See additional remarks under *inflatus*.

RANGE. Alaska, Yukon Territory, Montana and Colorado eastward across southern Canada and the eastern half of the United States Collections May through October.

HOST. Tenthredinidae: Nematinae: *Pikonema alaskensis* (Rohwer), yellowheaded spruce sawfly (UM) (Rau, 1976: 25-26, 33, 35-36; Houseweart and Kulman, 1976: 863-864; Thompson and Kulman 1980: 25-28); *Nematus tibialis* Newman, a willow sawfly (Brimley, 1942: 30); *Pristiphora erichsonii* (Hartig), larch sawfly (CNC). Diprionidae: *Diprion similis* (Hartig), introduced pine sawfly (Riley and Howard, 1890: 153); *Neodiprion sertifer* (Geoffroy), European pine sawfly (CNC).

MATERIAL EXAMINED. **Holotype**: male, QUEBEC, Provancher (UL). No paratypes. Other type material: *Al. pulchra* Provancher, male, QUEBEC, Provancher (UL); *Ph. pallicoxus* Provancher, male, QUEBEC, Provancher (UL). Other material studied: 845 M, 191 F (AEI, CAS, CNC, Dasch, INF, UCB, UCD, UCR, UM, USNM): ALASKA, ALBERTA, BRITISH COLUMBIA, COLORADO, MAINE, MANITOBA, MICHIGAN, MINNESOTA, MONTANA, NEW BRUNSWICK, NEWFOUNDLAND, NEW HAMPSHIRE, NEW JERSEY, NEW YORK, NORTH CAROLINA, OHIO, ONTARIO, PRINCE EDWARD ISLAND, QUEBEC,

SASKATCHEWAN, SOUTH CAROLINA, VERMONT, VIRGINIA, WEST VIRGINIA, WISCONSIN, YUKON TERRITORY

ETYMOLOGY. *Pubescens* (downy), probably referring to the fine, dense hairs of the head and thorax, although these are characteristic of the genus.

Endasys rotundiceps (Provancher)
(Fig. 68:A-D; Map 10)

Phygadeuon rotundiceps Provancher, 1877: 12; 1879: 74; 1882: 335; 1883: 320, 774;
   1886: 47. **Lectotype** designated by Gahan and Rohwer, 1918: 166, cited by Barron,
   1975: 545. Lists: Evans, 1896: 10; Dalla Torre, 1902: 693; Nason, 1905: 150;
   Cushman and Gahan 1921: 162; Johnson, 1927: 142.
Endasys (Endasys) rotundiceps: Townes, 1944: 214; Townes and Townes, 1951: 246.
Endasys rotundiceps: Carlson, 1979: 417.

MALE DIAGNOSIS. Generally medium-small, 5-6 mm long, sometimes smaller; white coxae I and II, sometimes coxa III apically, and trochanters; coxa III and femur III mostly orangish yellow, femur III and tibia III blackish apically; abdomen mostly brownish on terga 1-3 and apex, variably yellowish with brown apically on terga 4-6; 1st tergum moderately slender (Fig. 84:F), postpetiole square; frons with dense, fine pits; femur III moderately swollen (Fig. 79:C), ratio 0.23-0.24; propodeal carinae distinct but not coarse, area dentipara distinctly narrowed, apophysis distinct; differs from *praerotundiceps* by the mostly orangish yellow femur III and tibia III, more brownish anterior abdominal terga and yellowish posterior terga, lack of 3 rows of short and white spines on hind tibia, denser punctation of the frons, and the more slender 1st tergum and femur III; eastern half of United States and southern Canada.

MALE DESCRIPTION. *White*: Scape, clypeus, mandible, tegula, coxae I and II except coxa II basally, often coxa III ventro-apically. *Yellow*: Flagellum ventrally, coxa II basally, most of coxa III orangish yellow except sometimes with blackish, femora I and II, femur III orangish yellow except apically, tibiae except tibiae I and II dorsally and tibia III apically, variably on abdominal terga 1-3, at least basally on terga 4-6; pale forms with terga 3-6 mostly yellow. *Brownish*: Usually abdominal terga 1-4 and apex, at least apically on terga 4-6; dark forms with abdomen mostly brown. *Black*: Flagellum dorsally, sometimes coxa III basally, femur III and tibia III apically, hind tarsomeres. *Punctation* (Fig. 86:A-E): Face finely rugulose with very dense pits; frons with dense, fine pits; temple with moderately dense, indistinct pits; propleurum densely puncate to impunctate, fine, more or less indistinct pits; mesopleurum impunctate centrally, otherwise with mostly sparse, distinct pits, surface a little rugulose; 1st and 2nd terga mostly smooth and shiny. *Shape*: Clypeal margin not upturned (Fig. 78:A); flagellum moderately stout (Fig. 80:C), its segments squarish or a little longer than wide on apical half; areola widely hexagonal, area dentipara distinctly narrowed, apophysis sharp and distinct, propodeal carinae distinct; 1st tergum moderately slender (Fig. 84:F), postpetiole square, 2nd tergum regular; hind femur moderately swollen (Fig. 79:C), ratio 0.23-0.24; Rs straight (Fig. 82:B). *Other*: Wings lightly tinted, 5th abdominal sternum membranous medially, and glumes dense.

FEMALE DIAGNOSIS. Generally medium, 5-6.5 mm long, sometimes smaller; flagellum a little elongate, mostly linear to apex (Fig. 80:F), more or less tricolored with flagellar segments 5-9 white, basal segments yellow; coxae and trochanters whitish or pale yellow, femora I and II and tibiae I and II yellow with latter whitish dorsally, femur III and tibiae III orange, sometimes blackish apically, abdomen orange; 1st tergum moderately slender (Fig. 84:C), postpetiole moderately widened, about 1.7 times as wide as long; radial cell straight and longer than 2nd-discoidal cell; 3rd valvula moderately long, as long as basal 3 hind tarsomeres or longer; propodeal carinae fine but distinct, area dentipara distinctly narrowed, apophysis distinct and vertically projecting; differs from *praerotundiceps* by the tricolored flagellum, elongate basal 3 flagellar segments, the entirely orange abdomen, the whitish or pale coxae and trochanters, and the mostly orange femur III and tibia III.

FEMALE DESCRIPTION. *White*: Flagellar segments 5-9, usually coxae I and II ventrally, coxa III ventro-apically, trochanters. *Yellow*: Flagellar segments 1-4, scape, tegula, coxae I and II basally, coxa III more dorso-apically, sometimes all coxae more pale yellow ventrally, femora I and II, tibiae I and II except pale yellow dorsally. *Orange*: Clypeus, mandible, femur III and tibia III except sometimes apices, hind tarsomeres orangish, and abdomen except 3rd valvula. *Black*: Flagellum apical half, sometimes femur III and tibia III blackish apically, 3rd valvula. *Punctation* (Fig. 86:B,D-E): Face rugulose with very dense, fine pits; frons a little rugulose with very dense pits; temple with dense to moderately dense, more or less distinct pits; propleurum with dense to very dense, fine pits; mesopleurum with variably dense to sparse, distinct pits, surface rugulose. *Shape*: Clypeal margin sharp but not upturned (Fig. 78:B); flagellum slightly elongate, more or less linear to apex, basal 3 segments a little elongate (Fig. 80:F,I); areola broadly hexagonal, area dentipara distinctly narrowed, apophysis distinct, vertically projecting, propodeal carinae moderately distinct; 1st tergum moderately slender (Fig. 84:C), postpetiole moderately widened, about 1.7 times as wide as long; femur III moderately swollen, ratio 0.24-0.26; Rs straight (Fig. 82:B). *Other*: 3rd valvula moderately long, as long as basal 3 hind tarsomeres or longer; wings a little tinted.

RANGE. Eastern half of the United States and southern Canada from Minnesota eastward to Quebec and southward to Florida and Louisiana. Collections June through September.

REMARKS. The application of this name is more restricted than previously. See discussion under *praerotundiceps*. For affinity of *rotundiceps*, see discussion under *chrysoleptus*.

MATERIAL EXAMINED. **Lectotype**: female, QUEBEC, Provancher (UL). **Paralectotypes**: 3 females (UL), none conspecific, all *praerotundiceps* . Other material studied: 34 M, 33 F (AEI, CNC, Dasch, FLDA): FLORIDA, KANSAS, LOUISIANA, MARYLAND, NEW YORK, NORTH CAROLINA, OHIO, SOUTH CAROLINA, VIRGINIA.

ETYMOLOGY. *Rotundi-* (round) + *ceps* (head), referring to the apparent shape of the head, although characteristic of females of the genus.

Endasys subclavatus (Say)
(Fig. 52:A-D; Map 22)

Cryptus subclavatus Harris, 1835: 584. (Nomen nudum).

Cryptus subclavatus Say, 1836: 237-238; Leconte, 1859: 693; Dalla Torre, 1902: 591. (Type destroyed.)

Phygadeuon vulgaris Cresson, 1864: 310 (Holotype here determined); Provancher, 1882: 334 and 354; 1883: 776; 1886: 46; Lists: Beach, 1892; Slosson, 1897: 237; Ashmead, 1900: 568; Dalla Torre, 1902: 697; Nason, 1905: 150; Smith, 1909: 632; Viereck, (1916) 1917: 335-336.

Platylabus ruficornis Provancher, 1886: 38. (Lectotype designated by Gahan and Rohwer, 1918: 168; cited by Barron, 1975: 545); Bradley, 1903: 278, 282; Townes, 1944: 215; Barron, 1975: 549.

Medophron monticola Ashmead manuscript name, *in* Slossen, 1900: 319; syn. by Cushman, 1922: 18; as nomen nudum by Carlson, 1979: 418.

Phygadeuon subclavatus: Smith, 1909: 630.

Phygadeuon (Bachia?) brittoni Viereck (1916) 1917: 335-336; Townes, 1944: 215. (Holotype by monotypy.)

(Cryptus) Stylocryptus subclavatus Cushman and Gahan, 1921: 162. **Neotype** designation.

Stylocryptus subclavatus: Cushman, 1922: 28, 1928: 928. Lists: Johnson, 1927: 142; 1930: 99; Walley, 1931: 92; Brimley, 1938: 406.

Stylocryptus vulgaris: Cushman, 1925: 389; 1928: 928; Brimley, 1938: 406.

Endasys (Endasys) subclavatus: Townes, 1944: 214; Townes and Townes, 1951: 246; Walkley, 1967: 94.

Endasys subclavatus: Townes, 1970: 386; Carlson, 1979: 418. Biology: Short, 1959: 433, 435; 1978: 47, 211, 484-485; Price, 1970(a): 445, 448, 451-452; 1970(b): 456; 1971: 587, 589-592, 594; 1972(a): 191-194; 1972(b): 1005-1015; 1973: 684; 1974(a): 76-84; 1974(b): 109, 1975: 155, 163, 239, 306; Rau, 1976: 25; Drooz et al., 1977: 60-62.

MALE DIAGNOSIS. Large, 7-9 mm long; 1st abdominal tergum long and slender (Fig. 84:G), postpetiole longer than wide, dorsomedian carinae distinct to at least middle of postpetiole; femur III moderately swollen (Fig. 79:C), ratio 0.23-0.24, orange often with black apex; tibia III orange with black apex, hind tarsomeres pale yellow, abdomen orange except apically and sometimes basally; clypeus yellow, margin not upturned (Fig. 78:A), but sharp; flagellum moderately slender to moderately stout (Fig. 80:B,C), segments longer than wide or squarish on apical half, often entirely orange, tyloids (2) short (Fig. 87:D); eastern half of northern United States and southern Canada.

MALE DESCRIPTION. *White*: Tegula, coxae I and II except basally, coxae III apico-ventrally, trochanters. *Yellow*: Scape, clypeus, mandible, coxa I basally, femora I and II, tibiae I and II except pale yellow dorsally, hind tarsomeres pale yellowish. *Orange*: Flagellum at least ventrally, often entirely; coxa II basally, coxa III except ventro-apically;

femur III except sometimes apically, tibia III except usually apically. *Black*: Often flagellum dorsally, sometimes femur III apically, usually tibia III apically, abdomen apically and sometimes basally; sometimes coxa III blackish basally. *Punctation* (Fig. 86:A-B,D-E): Face finely rugulose with very dense, fine pits; frons with dense to moderately dense, distinct pits; temple with moderately dense to moderately sparse, more or less indistinct pits; propleurum densely punctate to impunctate, mostly smooth with more or less distinct pits; mesopleurum impunctate centrally, otherwise with mostly sparse, distinct pits, surface very slightly rugulose, mostly smooth and shiny; 1st and 2nd terga usually slightly mat. *Shape*: Clypeal margin not upturned (Fig. 78:A); flagellum moderately slender to moderately stout, its segments squarish or longer than wide on apical half; areola broadly to widely hexagonal, area dentipara regular or narrowed, apophysis distinct, propodeal carinae distinct; 1st tergum long and slender, postpetiole longer than wide or square, dorsomedian carina distinct beyond propodeal spiracle, 2nd tergum wide or regular; femur III moderately swollen (Fig. 79:C), ratio 0.23-0.24. *Other*: Tyloids (2) short (Fig. 87:D), wings tinted slightly yellowish, 5th abdominal sternum membranous medially, glumes moderately dense.

FEMALE DIAGNOSIS. Large, 6-8 mm long; clypeus black or orangish, margin very slightly upturned (Fig. 78:B); flagellum bicolored, moderately stout, linear to apex (Fig. 80:F), basal 3 segments short; femur III moderately swollen (Fig. 79:D), ratio 0.26-0.28; femur III and tibia III orange with black apices, hind tarsomeres orangish; 1st abdominal tergum moderately slender to moderately stout (Fig. 84:B), postpetiole moderately widened, about 1.7 times as wide as long, sides nearly parallel; 3rd valvula moderately long, about as long as basal 4 hind tarsomeres; propodeum level dorsally, carinae weak, areola broadly hexagonal, area dentipara narrowed and elongate apically, apophysis distinct and a little widened.

FEMALE DESCRIPTION. *Yellow*: Flagellum on basal half, scape, tegula, legs I and II, hind trochanters. *Orange*: Clypeus sometimes, mandible, coxa III except sometimes basally, femur III and tibia III except often apices, abdomen. *Black*: Flagellum on apical half, clypeus often, often femur III and tibia III apically, sometimes coxa III basally. *Punctation* (Fig. 86:B,D-F): Face finely rugulose with very dense pits; frons a little rugulose with very dense, coarse pits; temple with moderately dense to dense punctation, more or less distinct pits; propleurum with dense, fine pits, surface mostly smooth, some rugulosity on lower part; mesopleurum variably punctate with distinct pits, surface rugulose. *Shape*: Clypeal margin very slightly upturned (Fig. 78:B); flagellum moderately stout, linear to apex, basal 3 segments short (Fig. 80:F,H); areola broadly hexagonal, area dentipara narrowed, elongate, apophysis distinct and wide, propodeum level dorsally, carinae weak; 1st tergum moderately stout to moderately slender, postpetiole moderately widened, about 1.7 times as wide as long, sides nearly parallel; femur III moderately swollen (Fig. 79:D), ratio 0.26-0.28; Rs mostly straight, slightly curved apically. *Other*: 3rd valvula moderately long, about as long as basal 4 hind tarsomeres, wings tinted yellowish.

REMARKS. Many specimens labelled "*subclavatus*" in collections are misidentifications of *patulus* (Viereck). All the reared "*subclavatus*" in the CNC, USNM, and FDA collections are *patulus*. See additional discussion under *patulus* and *inflatus*. For Nearctic affinities of the species, see remarks under *hesperus* and *aurigena*. It is most closely related to Palearctic *annulatus* (Habermehl). Males differ by the lack of black on the 2d abdominal tergum and 2 tyloids, females by the lack of the white flagellar segments 5-9.

RANGE. Eastern half of northern United States and southern Canada. Collections spring through fall.

HOST. All reared specimens labelled "*subclavatus*" are *patulus*, so I have listed all published host records under it. Thus confirmed hosts of *subclavatus* are unknown.

TAXONOMIC REMARKS. The types of Thomas Say were destroyed. The original description (Say, 1836: 238) mentions 2 females: 1 from Pennsylvania and 1 from Massachusetts. From 1864-1921, only the synonyms *Platylabus ruficornis* Provancher (males) and *Phygadeuon vulgaris* Cresson (females) appear in the literature (except Smith, 1909). So it would appear that the type of *subclavatus* was destroyed at an early time. Only after Cushman and Gahan (1921) designated a neotype did *subclavatus* reappear in the literature.

From Say's original description, however, it is clear that he did not describe the species now called *subclavatus*. It is possible that it was really the same as Provancher's *rotundiceps* (old sense, not mine). Say clearly describes the species as small, under 0.20 inch, oval abdomen, front and middle legs honey-yellow, flagellum subclavate, and abdomen dusky or blackish at base and apex, all characteristic of *praerotundiceps* n. sp. ("old *rotundiceps*"). The neotype by contrast is over 0.25 inch long, legs are orangish, and abdomen is entirely orange. Cushman and Gahan disregarded the blackish in Say's description as simply variation in color. However, their "new *subclavatus*" never has black on the abdomen. Had the neotype never been designated, the correct name for "new *subclavatus*" would have been Cresson's *vulgaris*. Nevertheless, the publication of their neotype makes this discussion academic. Furthermore, it would no longer be feasible to change the name, since *subclavatus* is common in the literature because of its use in ecological studies.

MATERIAL EXAMINED. **Neotype**: female, QUEBEC (USNM). Other type material: *Ph. vulgaris* Cresson, female, ILLINOIS, 1864, Cresson (ANSP); *Pl. ruficornis* Provancher, male, QUEBEC, Provancher (UL). Other material studied: 464 M, 129 F (AEI, CAS, CNC, Dasch, MCZ, UCD, UM, USNM): ARKANSAS, CONNECTICUT, GEORGIA, ILLINOIS, IOWA, KANSAS, MAINE, MARYLAND, MASSACHUSETTS, MICHIGAN, MINNESOTA, MISSOURI, NEWFOUNDLAND, NEW JERSEY, NEW YORK, NORTH CAROLINA, OHIO, OKLAHOMA, ONTARIO, PENNSYLVANIA, QUEBEC, RHODE ISLAND, SOUTH CAROLINA, TENNESSEE, VERMONT, VIRGINIA, WASHINGTON D.C., WEST VIRGINIA.

ETYMOLOGY. *Subclavatus* (club-like), probably referring to the appearance of the female flagellum.

Endasys tyloidiphorus Luhman, sp. n.
(Fig. 38:A-D; Map 18)

MALE DIAGNOSIS. Medium large, 5.5-6.5 mm long; face very finely granular with very fine pits; clypeal margin more or less upturned (Fig.78:C), medially giving clypeus slightly beaked appearance; at least 3 short, prominent tyloids on flagellar segments 10-12 or 9-11, flagellum long and moderately slender to moderately stout, segments a little longer than wide on apical half; propodeum short, area dentipara narrowed, apophysis distinct; abdominal terga 2 and 3 orange and variably black, terga 4-6 orange, coxae I and II mostly white, trochanters white, coxa III mostly orange with white ventrally and apically, often some black basally, femur III orange with black apex, tibia III orange with blackish apically and dorsally; eastern half of United States and southern Canada.

MALE DESCRIPTION. *White*: Scape, clypeus, mandible, tegula, most of coxae I and II, coxa III ventrally and apically, trochanters. *Yellow*: Femora I and II, tibia I and II except pale yellow dorsally. *Orange*: Flagellum ventrally pale orange, coxa III variable basally, femur III except usually apex, tibia III except usually apically and variably dorsally, abdominal tergum 1 apically, variably on terga 2 and 3, terga 4-6 entirely, and terga 7-apex variable. *Black*: Flagellum except ventrally, often coxa III basally, femur III usually apically, tibia III usually apically and variably dorsally, hind tarsomeres; dark forms with terga 1-3 mostly blackish. *Punctation* (Fig. 86:A-D): Face very finely granular, pits very fine and very dense; frons with dense, fine pits; temple with moderately dense indistinct pits; propleurum with variably dense to sparse punctation, pits more or less distinct and variably rugulose; mesopleurum impunctate centrally and slightly rugulose, otherwise with variably sparse, more or less distinct pits and some rugulosity; 1st and 2nd terga smooth and shiny. *Shape*: Clypeal margin slightly upturned medially, clypeus often appearing beaked; flagellum long, moderately slender to moderately stout, its segments longer than wide or squarish on apical half; areola widely hexagonal, area dentipara distinctly narrowed, apophysis distinct, propodeum short, carinae distinct; 1st tergum moderately slender (Fig. 84:F), postpetiole longer than wide or squarish, 2nd tergum wide; femur III moderately swollen (Fig. 79:C), usually more basally than apically, ratio 0.23-0.24. *Other*: At least 3 short, prominent tyloids (Fig. 87:D) on flagellar segments 10-12 or 9-11, additional tyloids may be basal or distal; wings more or less hyaline; 5th abdominal sternum membranous or complete apically; glumes dense (Fig. 88:A).

FEMALE DIAGNOSIS. Medium large, 5.5-6.5 mm long; 3rd valvula very long, 1.25 to 1.5 times length of hind tarsomeres; face very finely rugulose with very dense, very fine pits; clypeus orange, margin sharp and upturned medially, often giving clypeus beaked appearance; legs I and II yellow, leg III mostly orange, sometimes coxa III black basally, abdomen orange except sometimes 1st tergum basally.

FEMALE DESCRIPTION. *Yellow*: Legs I and II, hind trochanters. *Orange*: Flagellum basal half, scape pale orange, clypeus blackish orange, mandible, tegula, leg III except trochanters and sometimes coxa basally, abdomen except sometimes 1st tergum basally. *Black*: Flagellum apical half, sometimes coxa III basally, sometimes 1st tergum basally. *Punctation* (Fig. 86:E-F): Face very finely rugulose with very dense, very fine pits; frons very finely rugulose with very dense pits; temple with moderately sparse, distinct pits; propleurum a little rugulose with very dense pits; mesopleurum with variably dense or sparse, distinct pits, surface rugulose. *Shape*: Clypeal margin sharp and upturned medially, often giving clypeus beaked appearance; flagellum moderately stout, nearly linear to apex, basal 3 segments short (Fig. 80:F,H); areola widely hexagonal, area dentipara distinctly narrowed, apophysis distinct, propodeum short with distinct carinae; 1st tergum moderately slender (Fig. 84:F), postpetiole gradually expanded, less than 1.5 times as wide as long; femur III moderately swollen or swollen, ratio 0.28-0.29; Rs straight. *Other*: 3rd valvula very long, 1.25-1.5 times longer than hind tarsomeres; wings more or less hyaline.

REMARKS. It appears related to *inflatus*, from which it differs by the male having 3 or more tyloids, and the female having a very long 3rd valvula.

RANGE. Eastern half of the United States and southern Canada. Collections April through August.

MATERIAL EXAMINED. **Holotype**: male, Ann Arbor, MICHIGAN, 22-vii-1959, H. and M. Townes (AEI). **Paratypes**: 72 M, 22 F (AEI, AMNH, CNC, Dasch, USNM, Yu): ALABAMA, ARKANSAS, ILLINOIS, IOWA, KANSAS, KENTUCKY, MARYLAND, MICHIGAN, MINNESOTA, MISSOURI, NEW JERSEY, NEW YORK, NORTH CAROLINA, OHIO, ONTARIO, PENNSYLVANIA, RHODE ISLAND, SOUTH CAROLINA, TENNESSEE, TEXAS, WEST VIRGINIA. Other material studied: 25 M (AEI, MCZ, UCD, UKS, UM): IOWA, MARYLAND, MASSACHUSETTS, MINNESOTA, OHIO, TEXAS, WEST VIRGINIA.

ETYMOLOGY. (Greek) *Tyl-* (knob) + *-oid-* (-like) + *phorus* (bearer), referring to the numerous (knob-like) tyloids of the male flagellum.

MONTICOLA GROUP

DIAGNOSIS. Clypeal margin not or weakly upturned, propodeal carinae distinct but not coarse, usually smooth between carinae, apophysis distinct or reduced, punctation usually fine, surfaces smooth and shiny, coxae and trochanters partly or entirely black, female flagellum usually orange or black, narrowed on apical half, basal 3 segments short or moniliform.

I have included 11 species in this group based largely on the shape of the clypeus, leg and abdominal color, and propodeal features. The following characteristics are common: clypeal margin not or weakly upturned, margin often distinct; face slightly wider than height of face plus clypeus; female flagellum mostly unicolored, short, and narrowed apically; punctation generally dense with fine pits, surfaces mostly smooth and shiny; propodeum long, slightly sloped, and shiny between distinct carinae, area dentipara narrowed mesally, widened laterally, apophysis more or less distinct; 1st abdominal tergum of male mostly moderately slender, dorsomedian carinae more or less distinct; hind femur of male mostly moderately slender; male 5th abdominal sternum complete or membranous medially, 6th never membranous; coxae and usually trochanters with black, never with white, hind leg of male mostly black, abdomen usually with black, often bicolored. The species of this group are generally found in western and northern North America; one species, *melanurus* (Roman), is Holarctic. A *Platycampus* species (Tenthredinidae) is the only record.

This group corresponds to the Palearctic Analis Group. The Nearctic species included in this group are *bicolorescens*, *brachyceratus*, *daschi*, *hexamerus*, *leioleptus*, *leopardus*, *melanurus* (Roman), *monticola* (Dalla Torre), *obscurus*, *pentacrocus*, and *serratus*.

Endasys bicolorescens Luhman, sp. n.
(Fig. 74:A-D; Map 2)

MALE DIAGNOSIS. Medium size, 5.5-7 mm long; clypeus black, margin sharp but not upturned (Fig. 78:B); 2 tyloids on flagellar segments 10 and 11; abdomen mostly black with orange usually on 1st tergum apically, 2nd apically, and 3rd basally; legs mostly black except trochantelli and 1st-trochanters at least apically yellowish; femur III slender to moderately slender (Fig. 79:B), ratio 0.20-0.22; wings hyaline; area dentipara nearly regular, widened laterally, apophysis more or less distinct; 1st tergum moderately stout or moderately slender, postpetiole a little longer than wide, 2nd tergum regular; flagellum moderately slender or moderately stout, segments longer than wide or square on apical half; British Columbia to Newfoundland and New Hampshire.

MALE DESCRIPTION. *Yellowish*: 1st trochanters at least apically, trochantelli, femur I more basally and apically, tibiae I and II except blackish apically; light forms with trochanters entirely yellowish, and legs I and II more yellowish. *Orange*: Sometimes 1st tergum apically, variably on 2nd tergum, usually apically, and usually 3rd tergum basally; light forms with mostly orangish scape, mandible, and abdomen more bicolored with orange on 1st tergum apically and most of terga 2-6. *Black*: Flagellum, scape, clypeus, mandible, tegula, coxae, 1st trochanters basally, femora except front femur more blackish yellow, tibia III and hind tarsomeres, most of tergum 1, variably on 2, and terga 3 apically to apex; dark forms with abdomen entirely black. *Punctation* (Fig. 86:A-E): Face finely rugulose with very dense pits; frons with dense, fine pits; temple with dense, indistinct pits; propleurum with dense to sparse punctation, fine, more or less distinct pits; mesopleurum with sparse to very sparse punctation centrally, otherwise with sparse, more or less distinct pits, surface slightly rugulose inferiorly; 1st and 2nd terga often very slightly mat and very slightly wrinkled, or sometimes smooth and shiny. *Shape*: Clypeal margin sharp but not upturned (Fig. 78:B); flagellum moderately slender to moderately stout, segments on distal half a little longer than wide or square; areola broadly hexagonal, area dentipara regular, widened laterally, apophysis more or less distinct, propodeal carinae distinct; 1st tergum moderately slender to moderately stout, postpetiole a little longer than wide, 2nd tergum regular; femur III slender to moderately slender, ratio 0.20-0.22; Rs straight (Fig. 82:B). *Other*: Wings hyaline, 5th abdominal sternum membranous medially, and glumes sparse.

FEMALE DIAGNOSIS. Medium small, about 5.5 mm long; yellowish coxae apically, trochanters, flagellum basal half, flagellar segments 5-9 paler; coxae basally and femora mostly black; abdomen more or less bicolored, black basally and on apical half, laterally on 2nd tergum, and apically on 3rd, remaining areas orange; femur III moderately slender (Fig. 79:B), ratio 0.23; flagellum moderately stout to moderately slender, more or less linear to apex, basal 3 segments elongate (Fig. 80:F,I); clypeus black, margin not upturned (Fig. 78:B); area dentipara distinctly narrowed, apophysis distinct, propodeum slightly sloped.

FEMALE DESCRIPTION. *Yellowish*: Flagellum basal half, segments 5-9 paler, scape, mandible, tegula, coxae apically, trochanters, femora I and II apically, tibiae I and II except apically. *Orange*: Tibia III at base, 1st abdominal tergum apically, 2nd except laterally, and 3rd except apically. *Black*: Flagellum apical half, clypeus, coxae basally, femora except femora I and II apically, tibia III except at base, hind tarsomeres, abdomen basally, 2nd tergum laterally, 3rd apically, terga 4-apex. *Punctation* (Fig. 86:A-D): Face and frons shiny, pits dense and more or less distinct; temple with dense, indistinct pits; propleurum finely rugulose with dense, fine, more or less distinct pits; mesopleurum with mostly sparse, fine pits, surface very slightly rugulose, otherwise smooth and shiny. *Shape*:

Clypeal margin evident but not upturned (Fig. 78:B); flagellum moderately stout or moderately slender, more or less linear to apex, basal 3 segments elongate (Fig. 80:F,I); areola broadly to widely hexagonal, area dentipara distinctly narrowed, a little elongate apically (Fig. 83:C), apophysis distinct, propodeum slightly sloped, carinae distinct; 1st tergum stout (Fig. 84:A), postpetiole distinctly widened, more than twice as wide as long; femur III moderately slender (Fig. 79:C), ratio about 0.23; radial cell a little swollen, Rs straight. *Other:* 3rd valvula moderately long, as long as basal 4 hind tarsomeres; wings slightly tinted.

REMARKS. This species may be related to *obscurus*. Males differ by the black scape and clypeus, abdominal terga 1-3 partly or entirely orange, larger size and stouter, 5.5-7 mm long, and its eastern distribution. Females differ by the basal 3 flagellar segments elongate, yellowish scape, tegula, and trochanters, and longer 3rd valvula, as long as the basal 4 hind tarsomeres. See additional discussion under *daschi*.

RANGE. British Columbia and Alberta eastward to Newfoundland and New Hampshire. Collections June through August.

MATERIAL EXAMINED. **Holotype**: male, Mt. Washington, 5100 ft., NEW HAMPSHIRE, 11-vii-1946, J. Peck and M. Townes (AEI). **Paratypes**: 35 M, 3 F (AEI, CNC, Dasch): ALBERTA, BRITISH COLUMBIA, NEW HAMPSHIRE, QUEBEC, YUKON TERRITORY.

ETYMOLOGY. *Bi-* (2) + *color* + *-escens* (becoming), referring to the more or less bicolored abdomen.

Endasys brachyceratus Luhman, sp. n.
(Fig. 48:A-D; Map 8)

MALE DIAGNOSIS. Medium, 5.5-6.5 mm long; flagellum short and stout (Fig. 80:D), 3 tyloids on flagellar segments 9-11; clypeus black without upturned margin, or only very slightly so (Fig. 78:A-B); all tibiae orange, coxae and 1st-trochanters black, trochantelli orange, femur III black; frons with variably dense, finely distinct pits; mesopleurum impunctate and rugulose centrally; area dentipara distinctly narrowed, apophysis and carinae strong; abdomen distinctly bicolored, black basally and on terga 3 apically to apex; Alaska and western Canada.

MALE DESCRIPTION. *Orange*: Scape, mandible, trochantelli, femora I and II, all tibiae, hind basitarsus except apically, abdominal terga 1 apically to 3 basally. *Black*: Flagellum, clypeus, tegula, coxae, 1st-trochanters more orangish black, femur III, hind tarsomeres except most of basitarsus, abdomen basally and terga 3 apically to apex; darker forms with black also on scape, mandible, and some blackish on femora I and II, and tibia III apically. *Punctation* (Fig. 86:B-E): Face rugulose with dense pits; frons with variably dense, finely distinct pits; temple with sparse, indistinct pits; propleurum impunctate or with dense, more or less indistinct pits; mesopleurum impunctate and rugulose centrally, otherwise rugulose with variable, more or less distinct pits; 1st and 2nd terga mat and slightly wrinkled. *Shape*: Clypeal margin very slightly upturned apically (Fig. 78:B); flagellum short and stout (Fig. 80:D), segments on distal half square or wider than long; areola widely hexagonal, area dentipara distinctly narrowed, apophysis distinct, propodeal carinae distinct; 1st tergum moderately stout (Fig. 84:E), postpetiole squarish, 2nd tergum regular to a little wide; femur III moderately slender (Fig. 79:B), ratio 0.21-0.22; Rs more or less straight. *Other*: Wings tinted yellowish, 5th abdominal sternum complete medially, and glumes sparse.

FEMALE DIAGNOSIS. Small to medium, 4.5-6.5 mm long; flagellum more or less bicolored, moderately slender, narrowed apically (Fig. 80:E); clypeus black, margin not upturned (Fig. 78:A); coxae black, trochanters orangish, femur III black, tibia III mostly orange, abdomen more or less bicolored, orange on terga 1 apically to 3 basally, remaining terga black; 3rd valvula long, as long as basal 4 hind tarsomeres; area dentipara narrowed, apophysis distinct; temple punctation moderately sparse with indistinct pits; Rs straight on basal 0.67, distinctly curved apical 0.33 (Fig. 82:C); 1st tergum moderately stout (Fig. 84:B), postpetiole abruptly widened, sides parallel.

FEMALE DESCRIPTION. *Orange*: Flagellum basal half, scape, mandible dark orange apically, trochanters orangish, femora I and II blackish orange, tibia III except often apically, hind tarsomeres, and abdominal terga 1 apically to 3 apically. *Black*: Flagellum mostly orangish apical half, clypeus, mandible except apically, tegula, coxae, femur III more orangish black, tibia III often a little apically, abdomen basally and on terga 3 apically to apex. *Punctation* (Fig. 86:B-E): Face slightly rugulose with dense pits; frons with dense, fine, more or less distinct pits; temple with moderately sparse, indistinct pits; propleurum with moderately dense, indistinct pits, surface smooth and shiny; mesopleurum with mostly sparse, fine, more or less distinct pits, surface a little rugulose. *Shape*: Clypeal margin not upturned (Fig. 78:A); flagellum moderately slender, narrowed apically, basal 3 segments short (Fig. 80:E,H); areola broadly to widely hexagonal, area dentipara distinctly narrowed, apophysis distinct, propodeal carinae distinct; 1st tergum moderately stout (Fig. 84:B), postpetiole abruptly widened, about twice as wide as long, sides nearly parallel; femur III moderately slender or moderately swollen (Fig. 79:C), ratio about 0.24; Rs straight on basal 0.67, distinctly curved apical 0.33 (Fig. 82:C). *Other*: 3rd valvula long, as long as basal 4 hind tarsomeres; wings a little tinted.

REMARKS. This species differs from all others in the Monticola Group by the 3 short tyloids on flagellar segments 9-11, and the tibia III entirely orange. Females differ from other species in this group by the combination of the bicolored flagellum, narrowed on apical half, basal 3 segments short, femur III black, and tibia III orange. Other species have either an orangish flagellum, or it is linear to the apex, or the basal 3 segments are moniliform.

RANGE. Alaska, Yukon Territory, and British Columbia. Collections June through August.

MATERIAL EXAMINED. **Holotype**: male, Racing River, 2400+ ft., BRITISH COLUMBIA, 27-vii-1973, H. and M. Townes (AEI). **Paratypes:** 7 M, 7 F (AEI, Dasch): ALASKA, BRITISH COLUMBIA, YUKON TERRITORY.

ETYMOLOGY. (Greek) *Brachy-* (short) + *ceratus* (horned), referring to the short, stout flagellum of the male.

Endasys daschi Luhman, sp. n.
(Fig. 19:A-D; Map 2)

MALE DIAGNOSIS. Medium small, 5-6.5 mm long; mostly black with orangish on abdominal terga 2 or 3 to 6, trochanters partly white, and femora I and II and tibiae I and II more yellowish black; punctation of head, thoracic pleura, and abdomen mostly smooth and shiny with fine to very fine pits; clypeus mostly convex, sometimes margin slightly upturned (Fig. 78:A); 1st abdominal tergum slender (Fig. 84:G), postpetiole elongate; area dentipara wide to regular, elongate laterally, apophysis distinct; flagellum moderately slender (Fig. 80:B), segments elongate; femur III moderately slender (Fig. 79:B), ratio 0.21-0.22; western United States.

MALE DESCRIPTION. *White*: 1st-trochanters at least apically, trochantelli entirely. *Yellow*: Femora I and II and tibiae I and II variably yellow to blackish; clypeus or scape sometimes blackish yellow; dark forms with trochanters more yellowish. *Orange*: Tibia III often at least basally, and sometimes more ventrally; 1st abdominal tergum often apically, terga 3-6 usually suffused with black, especially apically and laterally. *Black*: Flagellum, scape, clypeus, mandible, tegula, coxae, usually 1st-trochanters except apically, femora I and II variably, femur III, tibia III except usually at base and sometimes ventrally, hind tarsomeres, abdomen basally, apically, tergum 2 at least basally, and terga 3-6 variably, at least apically and laterally. *Punctation* (Fig. 86:A-C): Face rugulose with very dense, very fine pits; frons with moderately dense to moderately sparse punctation, fine pits; temple with moderately sparse, fine pits; propleurum impunctate or with dense, very fine, mostly indistinct pits, surface smooth and shiny; mesopleurum impunctate centrally, otherwise with evenly sparse, fine pits, surface smooth and shiny; 1st and 2nd terga smooth and shiny. *Shape*: Clypeus mostly convex, sometimes margin slightly upturned (Fig. 78:A); flagellum moderately slender (Fig. 80:B), segments longer than wide to apex; areola broadly to elongate hexagonal, area dentipara regular to wide, apophysis distinct, propodeal carinae distinct; 1st tergum slender, postpetiole elongate, 2nd tergum narrowed; femur III moderately slender (Fig. 79:B), ratio 0.20-0.21. *Other*: Wings more or less hyaline, 5th abdominal sternum membranous medially, glumes dense.

FEMALE DIAGNOSIS. Medium small, about 6 mm long; legs I and II yellowish-black, leg III with black coxa and femur, more orangish hind trochanters, tibia III, and hind tarsomeres; abdomen black basally and apically, and terga 3-6 apically and laterally, remaining areas orangish; flagellum blackish orange to blackish, moderately slender, a little narrowed apically (Fig. 80:E); 1st abdominal tergum moderately stout (Fig. 84:B), postpetiole moderately widened, about 1.5 times as wide as long; 3rd valvula moderately short, about as long as basal 3 hind tarsomeres; area dentipara narrowed, elongate apically, apophysis distinct; punctation of head generally finely indistinct; femur III moderately swollen (Fig. 79:D), ratio about 0.26.

FEMALE DESCRIPTION. *Blackish yellow*: Femora I and II apically, tibiae I and II. *Orange*: Sometimes scape; tibia III blackish orange, darker apically; abdomen mostly orangish on terga 3-6 basally, and often tergum 1 apically. *Black*: Flagellum mostly blackish, a little lighter on basal half ventrally, scape usually, clypeus, mandible, tegula, coxae, trochanters more yellowish black, femur III, tibia III apically, abdomen basally, apically, and terga 3-6 at least apically and laterally. *Punctation* (Fig. 86:A-D): Face finely granular with finely dense pits, frons with dense, variably distinct pits; temple with moderately dense, finely indistinct pits; propleurum with mostly moderately dense, very fine pits; surface smooth and shiny; mesopleurum with sparse, very fine pits, surface slightly rugulose but shiny. *Shape*: Clypeal margin not distinctly upturned (Fig. 78:A); flagellum moderately slender, a little narrowed apically, basal 3 segments short (Fig. 80:H); areola broadly hexagonal, area dentipara narrowed, elongate apically, apophysis distinct, propodeal carinae weak; 1st tergum moderately stout (Fig. 84:B), postpetiole moderately widened to about 1.5 times as wide as long; femur III moderately swollen (Fig. 79:D), ratio about 0.26; radial cell short, Rs curved apically (Fig. 82:C). *Other*: 3rd valvula moderately short, about as long as basal 3 hind tarsomeres; wings tinted.

REMARKS. This species may be related to *obscurus*. Males are distinguished by the narrow postpetiole, mostly blackish scape, whitish trochanters. Females differ by having the flagellum orangish or blackish, not bicolored, and slightly narrowed on apical half, and more swollen femur III (ratio 0.26). See additional remarks under *leioleptus*.

RANGE. Colorado, Idaho, and Montana. Collections July and August.

MATERIAL EXAMINED. **Holotype:** male, Doolittle Ranch, Mt. Evans, 9800 ft., COLORADO, 7-vii-1964, C. Dasch (Dasch). **Paratypes:** 89 M, 4 F (AEI, Dasch): COLORADO, IDAHO, MONTANA. Other material studied: 99 M (AEI, Dasch): COLORADO.

ETYMOLOGY. After Clement Dasch, who collected most of the specimens of this species studied.

Endasys hexamerus Luhman, sp. n.
(Fig. 25:A-D; Map 8)

MALE DIAGNOSIS. Medium, 5-7 mm long; mostly orange legs and abdomen, black coxae, tibia III apically, hind tarsomeres, and abdomen basally and apically; clypeus orangish black, margin sharp and very slightly upturned (Fig. 78:B); face finely granular with very dense pits, frons mostly smooth and shiny with fine, more or less distinct pits; flagellum moderately slender (Fig. 80:B), segments square on apical half; 1st abdominal tergum moderately stout (Fig. 84:E), postpetiole square, 1st and 2nd terga slightly mat and often very slightly wrinkled; femur III moderately slender (Fig. 79:B), ratio 0.21-0.22; wings distinctly blackish; Pacific Northwest to Wyoming, Alberta, and Saskatchewan.

MALE DESCRIPTION. *Orange*: Scape often, trochanters, legs I and II more yellowish orange, femur III except sometimes apically, tibia III except apically, and abdomen except basally and apically; pale forms with mostly orange scape, clypeus, and coxae. *Black*: Flagellum, clypeus except sometimes more orangish black, scape often, tegula, coxae, sometimes femur III apically, tibia III apically, hind tarsomeres, abdomen basally and apically; dark forms with black abdominal terga 4-6 apically. *Punctation* (Fig. 86:C-E): Face finely granular with very dense pits; frons with dense to moderately dense punctation, fine, more or less distinct pits; temple with dense, indistinct pits; propleurum with dense to sparse punctation, fine, more or less distinct pits; mesopleurum impunctate centrally, otherwise variably punctate with distinct pits, surface slightly rugulose; 1st and 2nd terga slightly mat and often very slightly wrinkled. *Shape*: Clypeal margin sharp, very slightly upturned (Fig. 78:B); flagellum moderately slender (Fig. 80:B), segments square or slightly longer than wide on apical half; areola widely hexagonal, area dentipara narrowed, apophysis distinct, propodeal carinae strong, propodeum very slightly sloped; 1st tergum moderately stout to moderately slender, postpetiole square or longer than wide; femur III moderately slender to moderately swollen, ratio 0.20-0.22. *Other*: Wings distinctly blackish, 5th abdominal sternum membranous medially, and glumes moderately dense to moderately sparse.

FEMALE DIAGNOSIS. Medium small to large, 5.5-7.5 mm long; flagellum orange, moderately slender, distinctly narrowed apically (Fig. 80:E), basal segment long, 2nd and 3rd short; orange legs, abdomen, scape, clypeus, and tegula; 1st tergum moderately stout (Fig. 84:B), postpetiole moderately wide, about 1.75 times as wide as long, sides parallel; 3rd valvula short, as long as basal 3 hind tarsomeres; femur III moderately swollen (Fig. 79:C), ratio 0.26; wings distinctly blackish; frons and temple densely, finely punctate, face rugulose with very dense pits.

FEMALE DESCRIPTION. *Orange*: Flagellum, scape, clypeus, mandible, tegula, legs, and abdomen. *Punctation* (Fig. 86:B,D): Face rugulose with very dense pits; frons with very dense, very finely distinct pits; temple with moderately dense, more or less distinct pits; propleurum smooth and shiny with moderately dense, very fine pits; mesopleurum rugulose with variable, more or less distinct pits; *Shape*: Clypeal margin sharp, slightly upturned (Fig. 78:B); flagellum moderately slender, distinctly narrowed apically (Fig. 80:E), basal segment long, 2nd and 3rd short; areola broadly hexagonal, area dentipara narrowed, apophysis distinct, propodeal carinae distinct; 1st tergum moderately stout, postpetiole moderately widened, more than 1.75 times as wide as long, sides parallel; femur III moderately swollen (Fig. 79:C), ratio about 0.26; Rs straight, very slightly bowed (Fig. 82:A). *Other*: 3rd valvula short, as long as basal 3 hind tarsomeres; wings distinctly blackish.

REMARKS. See discussion under *monticola*.

RANGE. Pacific Northwest from northern California to British Columbia, eastward to Saskatchewan, and Wyoming. Collections May through August.

MATERIAL EXAMINED. **Holotype**: male, Mt. Hood, OREGON, 30-vii-1978, H. and M. Townes (AEI). **Paratypes**: 57 M, 21 F (AEI, CAS, Dasch, OSU, UCB, UCR): ALBERTA, BRITISH COLUMBIA, CALIFORNIA, OREGON, SASKATCHEWAN, WASHINGTON, WYOMING. Other material studied: 27 M (AEI, CAS, Dasch, UCB, UCR, USNM): CALIFORNIA, OREGON.

ETYMOLOGY. (Greek) *Hexa-* (6) + *merus* (part), referring to the pattern of coloration of the legs and abdominal terga.

Endasys leioleptus Luhman, sp n.
(Fig. 60:A-D; Map 11)

MALE DIAGNOSIS. Medium small to large, 5-8 mm long; clypeus black, margin more or less distinctly upturned (Fig. 78:C); flagellum slender (Fig. 80:A), segments longer than wide to apex, apical segment distinctly narrowed; coxae and trochanters usually black, femora mostly orange except femur III black apically, tibiae mostly orange except tibia III basally and apically; femur III slender (Fig. 79:A), ratio 0.18-0.20; wings more or less hyaline, areolet a little enlarged; 1st tergum slender (Fig. 84:G), postpetiole distinctly narrowed, 2nd tergum narrowed; abdomen black at least basally and apically, and variably on terga 4 and 5; punctation of head and thoracic pleura generally very fine with more or less distinct pits, surfaces generally smooth and shiny; area dentipara nearly regular, apophysis more or less distinct; western United States and Canada.

MALE DESCRIPTION. *Orange*: Femora I and II and tibiae I and II more yellowish orange , femur III except apically, tibia III except basally and apically, most of abdominal terga 2-5 except often apically on terga 4 and 5; pale forms also with orange scape, clypeus, tegula, coxae, and trochanters. *Black*: Flagellum, scape, clypeus, mandible, tegula, coxae, trochanters except trochantelli more orange, femur III apically, tibia III basally and apically, hind tarsomeres; abdomen basally, apically, and often on terga 4 and 5 apically; small, dark forms with most of abdomen blackish. *Punctation* (Fig. 86:B-C): Face very finely granular with very dense pits; frons and propleurum with dense, very fine pits, surfaces smooth and shiny; temple with moderately dense, indistinct pits; mesopleurum with very sparsely punctate to impunctate centrally, otherwise with mostly sparse, very fine pits, surface smooth and shiny; 1st and 2nd terga smooth and shiny. *Shape*: Clypeal margin more or less distinctly upturned (Fig. 78:C); flagellum slender, segments longer than wide to apex, apical segment distinctly narrowed; areola irregularly hexagonal, often elongate apically (Fig. 85:B), area dentipara regular, apophysis more or less distinct, often slightly flared, propodeal carinae distinct; 1st tergum long and slender (Fig. 84:G), postpetiole distinctly narrowed, 2nd tergum narrowed; femur III slender (Fig. 79:A), ratio 0.18-0.20; Rs more or less straight. *Other*: Wings hyaline, 5th abdominal sternum membranous medially, glumes moderately dense, tyloids moderately long (Fig. 87:B), areolet a little enlarged.

FEMALE DIAGNOSIS. Small to medium, 4.5-6 mm long; face wide, more than 3 times wider than high, unevenly, coarsely punctate with some rugulosity, surface shiny; clypeus black, impressed apically, margin more or less distinctly upturned (Fig. 78:D); frons with dense punctation near antennae, sparse on vertex; flagellum orange, short and slender, narrowed apically (Fig. 80:E); legs mostly orangish black, abdomen usually blackish basally and on apical half, remaining terga orange; 3rd valvula moderately long, about as long as basal 4 hind tarsomeres; 1st tergum moderately stout (Fig. 84:B), postpetiole moderately widened, about 1.5 times as wide as long.

FEMALE DESCRIPTION. *Orange*: Flagellum, scape, mandible except blackish orange apically, femora more orangish black, tibiae, hind tarsomeres, abdomen on tergum 1 apically, tergum 2 entirely, and tergum 3 basally. *Black*: Clypeus, most of mandible, coxae, trochanters, femora more orangish black, and abdomen basally and on terga 3 apically to apex. *Punctation* (Fig. 86:A-B,D-E): Face rugulosity with uneven, coarse pits, surface shiny; frons with dense, fine pits on lower part, with very sparse, fine pits on vertex; temple with sparse, more or less indistinct pits; propleurum with moderately dense, fine, more or less indistinct pits; mesopleurum with variably sparse to very sparse, fine, more or less distinct pits, surface mostly smooth and shiny; metapleurum mostly smooth and shiny with very fine pits. *Shape*: Clypeus impressed apically, margin more or less distinctly upturned (Fig. 78:C); flagellum short and slender, narrowed apically, basal 3 segments short (Fig. 80:E,H); areola broadly hexagonal, area dentipara narrowed, apophysis distinct, propodeal carinae distinct but fine; 1st tergum moderately stout (Fig. 84:B), postpetiole moderately widened to about 1.5 times length; femur III moderately swollen (Fig. 79:D), ratio 0.27-0.28; Rs straight except at apex (Fig. 82:C). *Other*: 3rd valvula moderately long, about as long as basal 4 hind tarsomeres; wings slightly tinted.

REMARKS. This species appears related to *daschi*. Males are distinguished by the orange trochantelli, not yellowish, the yellowish orange hind femur and tibia with black apically, and the mostly orange abdominal terga 2-6, the anterior terga never completely black. Female *leioleptus* differ by the orange, distinctly narrowed, short flagellum, and the abdomen weakly bicolored, without black anteriorly.

RANGE. Pacific Northwest from British Columbia to northern California eastward to Idaho and Colorado. Collections April, and mostly June through August.

HOST. Tenthredinidae, Tenthredininae: *Platycampus* sp., sawfly on *Populus* or *Larix*; another series of *leioleptus* was collected on *Populus trichocarpa* (USNM).

MATERIAL EXAMINED. **Holotype**: male, Mt. Ranier, 5500 ft., WASHINGTON, 23-vii-1940, H. and M. Townes (AEI). **Paratypes**: 64 M, 12 F (AEI, CAS, Dasch, OSU, USNM): CALIFORNIA, COLORADO, IDAHO, OREGON, WASHINGTON. Other material studied: 66 M, 3 F (AEI, CAS, Dasch, OSU): BRITISH COLUMBIA, CALIFORNIA, COLORADO, IDAHO, OREGON, WASHINGTON.

ETYMOLOGY. (Greek) *Leio-* (smooth) + *leptus* (slender), referring to the slender, shiny legs and body.

Endasys leopardus Luhman, sp. n.
(Fig. 46:A-D; Map 10)

MALE DIAGNOSIS. Medium, 5.5-6.5 mm long; mostly yellowish orange legs with coxae black basally, leg III mostly black with yellowish orange coxa apically and trochanters; abdomen more or less bicolored, orange mostly 2nd tergum and basally on 3rd, often blackish markings on 2nd tergum, remaining terga black; clypeus orange without upturned margin (Fig. 78:A); flagellum moderately stout (Fig. 80:C), segments squarish to apex, glumes sparse, usually orange around bases of tyloids (2); propodeum sloped, area dentipara nearly regular, widened laterally, apophysis more or less distinct; 1st tergum moderately slender (Fig. 84:F), postpetiole a little longer than wide; femur III moderately slender (Fig. 79:B), ratio 0.21 0.22; Alaska and Canada.

MALE DESCRIPTION. *Orange*: Scape, clypeus, mandible, area around tyloid bases, tegula, coxae apically, trochanters, femora I and II, tibiae I and II except often with blackish, femur III variable, tibia III basally, and abdomen usually on 1st tergum apically, variably on 2nd, 3rd apically, and 4th to apex. *Black*: Most of flagellum, coxae basally, femur II and tibia II often with blackish, femur III variably, tibia III except basally, hind tarsomeres, most of abdomen except usually 1st tergum apically, 2nd variably, and 3rd basally; smaller forms with mostly black head, legs, and abdomen except orangish on terga 2 and 3. *Punctation* (Fig. 86:B-D): Face finely granular or finely rugulose with very dense, fine pits; frons with dense, finely distinct pits; temple with dense, indistinct pits; propleurum with dense to sparse punctation, fine, more or less distinct pits; mesopleurum impunctate centrally, otherwise with mostly sparse, finely distinct pits, surface slightly rugulose; 1st and 2nd terga more or less mat. *Shape*: Clypeus without upturned margin (Fig. 78:A); flagellum moderately stout (Fig. 80:C), segments squarish on apical half; areola broadly hexagonal, area dentipara regular, widened laterally, apophysis moderately strong, propodeum a little sloped, carinae weak or moderately distinct; 1st tergum moderately slender (Fig. 84:F), postpetiole longer than wide, 2nd tergum regular; femur III moderately slender (Fig. 79:B), ratio 0.20-0.21; radial cell long, Rs more or less straight. *Other*: Wings more or less hyaline or tinted; 5th abdominal sternum complete medially; glumes sparse.

FEMALE DIAGNOSIS. Medium, 5.5-6.5 mm long; flagellum orangish, short and stout, linear to apex (Fig. 80:F), basal 3 segments nearly moniliform; clypeus orange without upturned margin (Fig. 78:A); face finely granular or finely rugulose with very dense, fine pits, temple densely punctate with indistinct pits; legs mostly orange with blackish coxae basally, variably on femora II and III, and apically on tibiae II and III; abdomen weakly bicolored, orange 1st and 2nd terga, and remaining terga except apically blackish; area dentipara narrowed, elongate apically, apophysis distinct; 1st tergum stout, postpetiole wide, about twice as wide as long; radial cell long, Rs straight on basal 0.75, curved on apical 0.25.

FEMALE DESCRIPTION. *Orange*: Flagellum dark orange, scape, clypeus usually, mandible, tegula, coxae apically, trochanters, femur I, tibia I, variably on femora II and III, tibiae II and III except apically, abdominal terga 1 and 2 entirely, and remaining terga except apically. *Black*: Coxae basally blackish, femora II and III variably, tibiae II and III apically, and apically terga 3-apex. *Punctation* (Fig. 86:B-D): Face finely rugulose to finely granular with very dense, fine pits; frons with dense, fine pits; temple and propleurum with dense, indistinct pits; mesopleurum with mostly sparse, fine, more or less distinct pits, surface mostly shiny with some rugulosity. *Shape*: Clypeus without upturned margin (Fig. 78:A); flagellum short and stout, linear to apex, basal 3 segments nearly moniliform (Fig. 80:F-G); areola boadly hexagonal with weak carinae, area dentipara narrowed, elongate apically (Fig. 83:C), apophysis distinct, carinae distinct; 1st tergum stout (Fig. 84:A), postpetiole wide, about twice as wide as long; femur III moderately swollen (Fig. 79:D), ratio 0.27-0.28; radial cell long, Rs basal 0.75 straight, apical 0.25 curved. *Other*: 3rd valvula moderately long, about as long as basal 4 hind tarsomeres; wings slightly tinted.

REMARKS. This species appears closely related to *obscurus* from which darker forms are difficult to distinguish. Male *leopardus* differs by the narrow postpetiole, more orangish scape, clypeus, front and middle coxae and trochanters, by orangish near the base of the tyloids, and the abdomen more bicolored, with less black on terga 2 and 3. Females are distinguished by the orangish flagellum, basal 3 segments moniliform, orange scape, clypeus, and trochanters, and the more swollen hind femur (ratio 0.27-0.28).

RANGE. Northern North America in Alaska, Yukon Territory, British Columbia, and Newfoundland. Collections June through August.

MATERIAL EXAMINED. **Holotype**: male, Stone Mt. Pk., 3800+ ft., BRITISH COLUMBIA, 17-vii-1973, H. and M. Townes (AEI). **Paratypes**: 49 M, 12 F (AEI, CNC, Dasch): ALASKA, ALBERTA, BRITISH COLUMBIA, NEWFOUNDLAND, YUKON TERRITORY. Other material studied: 141 M, 18 F (AEI, INF): ALASKA, BRITISH COLUMBIA, IDAHO, YUKON TERRITORY.

ETYMOLOGY. *Leopardus* (leopard), referring to the blackish spots often on terga 2 and 3 of the male.

Endasys melanurus (Roman)
(Fig. 27:A-D; Map 15)

Stylocryptus melanurus Roman, 1909: 243, 1913: 122; Habermehl, 1916: 379; Schmiedeknecht, 1933: 51. **Lectotype** designated by Sawoniewicz, 1981 (Luhman and Sawoniewicz, 1991.)
Stylocryptus analis: Roman, 1909: 243 (misidentified).
Stylocryptus bicolor var. *melanurus:* Schmiedeknecht, 1933: 23, 29, 52.
Endasys bicolor melanurus: Jussila, 1973: 19.
Endasys melanurus Luhman and Sawoniewicz,1991.

MALE DIAGNOSIS. Medium, 4.5-6 mm long; abdomen mostly black except often basal half of tergum 3 orangish, sometimes variably on tergum 2; tibia III pale orange except apically; hairs more or less erect giving body a fuzzy appearance; flagellum stout (Fig. 80:D), segments square to wide on apical 3rd; clypeal margin not upturned; punctation of face and frons a little rugulose with variably dense pits; punctation of temple, propleurum, and mesopleurum with mostly sparse, more or less distinct pits; Norway, Sweden, Alaska, and northern Canada.

MALE DESCRIPTION. *Pale orange*: Femur I apically, tibiae I and II, tibia III except apically, sometimes 3rd abdominal tergum on basal half. *Black*: Flagellum, scape, mandible, tegula, legs I and II except femur I apically, leg III except tibia III black apically, most of abdomen, except 2nd tergum blackish orange and 3rd tergum orange on basal half, sometimes abdomen entirely black. *Punctation* (Fig. 86:B,D): Face slightly rugulose with variably dense punctation; frons with dense, fine pits; temple with sparse, more or less distinct pits; propleurum with sparse, fine pits; mesopleurum with sparse, more or less distinct pits, surface variably rugulose; 1st and 2nd abdominal terga slightly wrinkled but shiny. *Shape*: Clypeal margin evident but not upturned (Fig. 78:B); flagellum stout (Fig. 80:D), segments square to wide on apical 0.67; areola nearly hexagonal, area dentipara regular or slightly wide, apophysis moderately strong, propodeal carinae weak; 1st tergum moderately stout (Fig. 84:E), postpetiole square, 2nd tergum widened; femur III moderately slender (Fig. 79:B), ratio about 0.22; radial cell long, about 1.5 times as long as 2nd discoidal cell, Rs straight. *Other*: Sometimes a 3rd, very small tyloid on flagellar segment 9 or 11; wings hyaline to slightly tinted; 5th abdominal sternum complete medially; hairs on head, legs, and thorax nearly erect giving body fuzzy appearance.

FEMALE DIAGNOSIS. Small size, 4.5-5.5 mm long; scape and clypeus black, face and clypeus more or less swollen, clypeus without upturned margin (Fig. 78:A), punctation of face variably dense to very sparse with variably sized pits, surface varying from rugulose to smooth and shiny, legs mostly black except pale orange or yellowish on tibiae and apically on femora I and II, abdomen bicolored on apical half, basal half mostly pale orange, flagellum bicolored and very short, basal 3 segments moniliform, 1st abdominal tergum stout (Fig. 84:D), postpetiole about as long as petiole.

FEMALE DESCRIPTION. *Yellowish*: Flagellum blackish yellow on basal half, femora I and II apically, tibiae and tarsomeres of legs I and II. *Pale orange*: Tibia and tarsomeres, 1st tergum apically, terga 2 and 3 entirely, and 4th basally. *Black*: Flagellum apical half, scape, clypeus, mandible, tegula, legs except femora I and II apically, tibiae I and II, tarsi I and II, abdomen on apical half from tergum 4 apically to apex, and most of tergum 1. *Punctation* (Fig. 86:B-C,E): Face rugulose or smooth and shiny with variably dense to very sparse pits of variable size; frons with dense, distinct pits; temple with sparse, indistinct pits; propleurum with sparse, fine pits; mesopleurum with sparse, more or less distinct pits, surface slightly rugulose; metapleurum finely rugulose with dense, fine pits. *Shape*: Clypeus swollen, margin not upturned (Fig. 78:A); flagellum short, moderately slender, narrowed apically, basal 3 segments nearly moniliform (Fig. 80:E,G); areola nearly hexagonal, area dentipara narrowed, apophysis weak, propodeal carinae weak to very weak; 1st tergum stout and wide (Fig. 84:A), postpetiole distinctly widened, about as long as long petiole; femur III moderately swollen (Fig. 79:D), ratio 0.27-0.28; Rs straight to apical 0.25, then distinctly curved (Fig. 82:D). *Other*: 3rd valvula moderately short, about as long as basal 3 hind tarsomeres; wings tinted; hairs on head, legs, and thorax nearly erect, giving body fuzzy appearance.

REMARKS. This Holarctic species is distinctive by the sparse, erect hairs of the body and appendages, giving it a fuzzy appearance. It does not appear closely related to other Nearctic species. Its closest relative may be found in the Palearctic. It superficially resembles *bicolor*, but differs by having a broadly oval clypeus, less than 3 times wider than high, and the clypeal margin not upturned.

RANGE. Holarctic, from Sweden, Norway, Alaska, Yukon Territory, British Columbia, and northern Quebec. Collections July and August.

MATERIAL EXAMINED. **Lectotype**: female, Sarek Mts., SWEDEN, 25-vii-1907, Roman (NRS). **Paralectotypes**: 1 M, 2 F (NRS): SWEDEN. Other material studied: 5 M, 2 F (AEI, CNC, Dasch, MCZ): ALASKA, BRITISH COLUMBIA, MASSACHUSETTS, QUEBEC, YUKON TERRITORY.

ETYMOLOGY. (Greek) *Melan-* (black) + *urus* (tail), referring to the black abdomen apically.

Endasys monticola (Dalla Torre)
(Fig. 9:A-D; Map 13)

Phygadeuon montanus Cresson, 1864: 308-309 (preoccupied according to Art. 59 [b] of the Code by *Phygadeuon montanus* [Gravenhorst, 1829: 616] in Schmiedeknecht, 1890: 150); Cresson, 1865: 265. **Holotype** by monotypy, here determined.
Phygadeuon monticola Dalla Torre, 1902: 689 (replacement name).
Endasys (Endasys) montanus: Townes, 1944: 212; Townes and Townes, 1951: 246.
Endasys monticola Carlson, 1979: 417.

MALE DIAGNOSIS. Medium large, 5.5-7 mm long; mostly black with orange abdominal terga 2 and 3, sometimes also trochantelli, femora, and tibiae more orangish; face shiny with very fine punctation; clypeal margin sharp and more or less distinctly upturned (Fig. 78: C); punctation of head and thoracic pleura generally sparse and very fine, surfaces smooth and shiny; mesopleurum sparsely punctate centrally; propodeum shiny, slightly depressed, carinae distinct, areola broadly to nearly hexagonal, area dentipara regular to narrowed, apophysis distinct; postpetiole and 2nd tergum slightly mat with very fine but distinct pits, surface a little rough; wings hyaline to slightly tinted; metapleurum shiny and smooth; western United States and Canada.

MALE DESCRIPTION. *Orange*: Femur I and sometimes II, at least apically; tibia I and sometimes II; sometimes tibia III more orangish basally; usually postpetiole apically and terga 2 and 3; light forms more orangish scape, mandible, tegula, legs, and terga 4-6. *Black*: Head, thorax, abdomen except for yellowish on leg II and orange on abdominal terga 1-3. *Punctation* (Fig. 86:B-D): Face shiny with very fine, indistinct pits; frons with moderately dense to sparse punctation, very fine, more or less distinct pits; temple with moderately dense, more or less indistinct pits; propleurum sparsely punctate to nearly impunctate, surface shiny and smooth; mesopleurum with variably sparse, very fine, indistinct pits, usually some pits on central area; metapleurum smooth and shiny, punctation indistinct; postpetiole and 2nd tergum slightly mat and rough, pits very fine but distinct. *Shape*: Clypeal margin sharp and more or less upturned (Fig. 78:C); flagellum moderately slender (Fig. 80:B), segments square or a little longer than wide on apical half of flagellum; areola broadly to nearly hexagonal, area dentipara regular to narrowed and slightly elongate, apophysis distinct, propodeum shiny between distinct carinae, slightly depressed apically; 1st tergum moderately slender (Fig. 84:F), postpetiole longer than wide; femur III moderately slender (Fig. 79:B), ratio about 0.20. *Other*: Wings hyaline to tinted, 5th abdominal sternum membranous medially, glumes moderately dense, and hind tibial spines in about 3 rows.

FEMALE DIAGNOSIS. Medium large, 6-6.5 mm long; clypeal margin more or less distinctly upturned (Fig. 78:C); flagellum orange, moderately slender (Fig. 80:E), narrowed apically; 3rd valvula short, less than basal 3 hind tarsomeres; 1st abdominal tergum moderately stout (Fig. 84:B), postpetiole moderately widened, about 1.75 times as wide as long, sides nearly parallel; Rs mostly straight, radial cell about as long as 2nd-discoidal cell; legs orange except some blackish on femur III and basally on coxae II and III; abdomen blackish laterally and apically on terga 3-7; face slightly rugulose with dense, variably sized pits; punctation generally very fine, denser on head, sparser on thoracic pleura, surfaces smooth and shiny; propodeum nearly level, generally smooth between dorsal carinae.

FEMALE DESCRIPTION. *Orange*: Flagellum, scape, mandible, tegula, coxae except coxae II and III basally, trochanters, legs except femur III blackish, and abdomen except terga 3-7 laterally and apically. *Black*: Coxae II and III at least basally, femur III variably, abdominal terga 3-7 laterally and apically. *Punctation* (Fig. 86:B-D): Face slightly rugulose with very dense, variably sized pits; frons with very dense, fine, more or less distinct pits; temple with moderately dense, more or less distinct pits; propleurum smooth and shiny with very fine, sparse pits; mesopleurum smooth and shiny with mostly sparse, very fine pits; metapleurum shiny with indistinct pits. *Shape*: Clypeal margin more or less distinctly upturned (Fig. 78:C); flagellum moderately slender, narrowed apically, basal 3 segments longer than wide (Fig. 80:E,H); areola broadly hexagonal, area dentipara narrowed, apophysis distinct, propodeum nearly level; 1st abdominal tergum moderately stout (Fig. 84:B), postpetiole moderately, but abruptly widened, about 1.75 times as wide as long; femur III moderately swollen (Fig. 79:D), ratio about 0.28; radial cell about as long as 2nd-discoidal cell. *Other*: 3rd valvula short, less than the length of basal 3 hind tarsomeres; wings tinted.

REMARKS. This species is closely related to *hexamerus* and *pentacrocus*. Male *monticola* is distinguished from both by the mesopleurum sparsely punctate centrally; additionally, it differs from the former by the black trochanters and femur III, and abdominal terga 3-6 partly black. It differs from *pentacrocus* by abdominal terga 3-6 partly or entirely black, trochantelli black, and narrow postpetiole with finely distinct pits. Female *monticola* differs from *hexamerus* by the black clypeus, coxa III blackish orange, and abdomen partly black apically; from *pentacrocus* by the orange flagellum narrowed on the apical half, the orange femur III, tibia III, and coxae I and II more orangish than blackish. See additional remarks under *serratus*.

RANGE. Pacific Northwest from northern California to Washington, westward to Alberta, Montana, Idaho, Wyoming, and Colorado. Collections June through August.

MATERIAL EXAMINED. **Holotype**: female, Rocky Mts, COLORADO, 1864, Cresson (ANSP). No paratypes. Other material studied: 160 M. 2 F (AEI, CAS, Dasch, MCZ, OSU, UCB, UCD, UCR): ALBERTA, CALIFORNIA, IDAHO, MONTANA, NEVADA, OREGON, SASKATCHEWAN, WASHINGTON, WYOMING.

ETYMOLOGY. *Montanus* (mountain dweller); *monti-* (mountain) + *cola* (dweller; masculine noun), referring to the habitat of the type.

Endasys obscurus Luhman, sp. n.
(Fig. 72:A-D; Map 11)

MALE DIAGNOSIS. Small, 4.5-5.5 mm long; abdomen mostly black with yellowish on terga 2 and 3, coxae and leg III black, trochanters mostly yellow; clypeus black, margin not upturned; face shiny, finely granular; femur III slender to moderately slender, ratio, 0.19-0.21; tyloids (2) moderately long and prominent; propodeal carinae moderately distinct, area dentipara nearly regular, apophysis more or less distinct; 1st tergum moderately stout, postpetiole squarish (Fig. 84:E); Rs more or less straight; 5th abdominal sternum complete medially; Alaska, British Columbia, and Oregon.

MALE DESCRIPTION. *Dark yellowish*: Scape, clypeus sometimes, flagellum baso-ventrally, trochanters except basally, most of femur I, femur II basally and apically, tibia I, tibiae II and III basally, abdomen on 1st tergum apically, 2nd variably, and 3rd basally. *Dark orangish*: Sometimes abdominal terga 1 apically and variably on 2 and 3. *Black*: Most of flagellum, usually clypeus, mandible, tegula, coxae, trochanters more basally, femur I blackish dorsally and ventrally, femur II basally and apically, femur III, tibiae II and III apically, hind tarsomeres, most of abdomen except 1st tergum apically, variably on terga 2 and 3. *Punctation* (Fig. 86:A-D): Face finely granular with dense, fine pits, surface shiny; frons with moderately dense, fine, more or less distinct pits; temple with moderately dense, indistinct pits; propleurum mostly impunctate, surface smooth and shiny; mesopleurum mostly impunctate centrally, otherwise with sparse, very fine, indistinct pits, surface slightly rugulose; metapleurum rugulose; 1st and 2nd terga slightly mat. *Shape*: Clypeal margin not upturned (Fig. 78:A); flagellum moderately stout (Fig. 80:C), its segments squarish on apical half; areola broadly hexagonal, area dentipara regular, apophysis weak but more or less distinct, propodeal carinae distinct; 1st tergum moderately stout (Fig. 84:E), postpetiole squarish or slightly longer than wide, 2nd tergum regular or wide; femur III slender to moderately slender, ratio 0.19-0.21; Rs more or less straight. *Other*: Sometimes 3rd tyloid on 9th flagellar segment, distal 2 moderately long (Fig. 87:B) and prominent; wings hyaline; 5th abdominal tergum complete; glumes sparse to very sparse at base (Fig. 88:C).

FEMALE DIAGNOSIS. Small, 4.5 mm long; abdomen mostly brownish with yellowish on 1st tergum apically and most of 2nd tergum, legs mostly blackish except trochanters and tibiae; Rs more or less straight; 1st tergum stout with postpetiole wide (Fig. 84:A); area dentipara narrowed and elongate apically, apophysis distinct; femur III moderately swollen (Fig. 79:C), ratio about 0.23; clypeus slightly swollen, margin not upturned (Fig. 78:A); flagellum bicolored, moderately stout, linear to apex (Fig. 80:F); frons densely and finely punctate, temple moderately punctate with indistinct pits.

FEMALE DESCRIPTION (based on one specimen). *Yellow*: Flagellum basal half, trochanters, femora I and II apically, tibiae I and II, tibia III except apically, and 1st tergum apically and 2nd except laterally. *Black*: Flagellum basal half, scape more brownish, clypeus, tegula, coxae, femora I and II except apically, femur III, tibia III apically, and most of abdomen except yellowish on terga 1 and 2. *Punctation* (Fig. 86:A-C): Face shiny with dense pits; frons with dense, fine pits; temple with sparse, indistinct pits; propleurum mostly smooth and shiny, hairs moderately densely spaced; mesopleurum with sparse, very fine pits. *Shape*: Clypeus slightly swollen, margin not upturned (Fig. 78:A); flagellum moderately stout, linear to apex, basal 3 segments slightly longer than wide (Fig. 80:F,H); areola broadly hexagonal, area dentipara narrowed, apophysis distinct, propodeal carinae fine but distinct; 1st tergum stout (Fig. 84:A), postpetiole distinctly widened; femur III moderately swollen (Fig. 79:C), ratio about 0.23; Rs more or less straight. *Other*: 3rd valvula moderately short, as long as basal 3 hind tarsomeres; wings slightly tinted.

REMARKS. It appears related to a new Palearctic species. For Nearctic relationships, see remarks under *daschi, leopardus,* and *bicolorescens.*

RANGE. Alaska, Yukon Territory, British Columbia, and Oregon. Collections late June through August.

MATERIAL EXAMINED. **Holotype**: male, Tsaina R., ALASKA, 16-viii-1973, H. and M. Townes (AEI). **Paratypes**: 27 M, 1 F (AEI, Dasch): ALASKA, BRITISH COLUMBIA, OREGON, YUKON TERRITORY. Other material studied: 11 M, 1 F (AEI, Dasch): ALASKA, BRITISH COLUMBIA, YUKON TERRITORY.

ETYMOLOGY. *Obscurus* (indistinct), referring to the obscure patterns of color and punctation, and the difficulty in characterizing the species.

Endasys pentacrocus Luhman, sp. n.
(Fig. 35:A-D; Map 18)

MALE DIAGNOSIS. Medium large, 6-7 mm long; face finely granular or rugulose with dense pits; clypeus black, margin slightly upturned (Fig. 78:B); flagellum moderately stout (Fig. 80:C), segments square or wider than long on apical half of flagellum, tyloids yellow near bases, glumes dense; black coxae, 1st-trochanters, femur III, and tibia III at least apically and basally; abdominal terga 2-6 mostly orange, black basally and apically; wings hyaline; 1st tergum moderately slender (Fig. 84:F), postpetiole a little longer than wide; area dentipara narrowed, apophysis distinct; punctation of head and thoracic pleura generally fine but distinct; 5th abdominal sternum complete medially; western United States and Canada.

MALE DESCRIPTION. *Yellow*: Basal area around tyloids, trochantelli I and II, femora I and II, tibiae I and II except often more yellowish black. *Orange*: Scape sometimes, hind trochantellus, femur III often basally, tibia III except at least basally and apically, abdomen except basally and apically. *Black*: Flagellum, scape usually, clypeus, mandible, tegula, coxae, 1st-trochanters, femora I and II and tibiae I and II often more yellowish black, femur III except often basally, tibia III at least basally and apically, hind tarsomeres, abdomen basally and apically, usually some blackish laterally and apically on terga 3-6; dark forms with increasing black on terga 2-6 apically and laterally. *Punctation* (Fig. 86:B-C,E): Face finely rugulose or granular with dense, fine pits; frons with moderately dense to moderately sparse punctation, very fine pits; temple with moderately sparse, indistinct pits; propleurum with dense, very fine pits, to impunctate; mesopleurum impunctate centrally, otherwise with sparse, distinct pits, surface a little rugulose; 1st and 2nd terga very slightly mat. *Shape*: Clypeal margin very slightly upturned (Fig. 78:B); flagellum moderately stout (Fig. 80:C), its segments square or wider than long on apical half; areola broadly hexagonal, area dentipara distinctly narrowed, apophysis distinct, propodeal carinae distinct; 1st tergum moderately slender (Fig. 84:F), postpetiole a little longer than wide, 2nd tergum a little widened; femur III moderately slender (Fig. 79:B), ratio 0.21-0.22; Rs straight. *Other*: Wings hyaline, 5th abdominal sternum complete medially, and glumes dense.

FEMALE DIAGNOSIS. Medium, 5.5-7 mm long; flagellum bicolored, linear to apex (Fig. 80:F); face finely rugulose with finely dense pits; clypeus black, margin not upturned; coxae and femora mostly black, trochanters blackish yellow, tibia III black, abdomen bicolored, mostly orange on terga 2 and 3, black on terga 4-apex and most of tergum 1; area dentipara narrowed, apophysis sharply distinct, propodeum level dorsally; 1st tergum moderately stout (Fig. 84:B), postpetiole moderately widened; femur III moderately swollen, ratio about 0.23; 3rd valvula moderately short, about as long as basal 3 hind tarsomeres; Rs more or less straight, wings tinted.

FEMALE DESCRIPTION. *Yellow*: Trochanters blackish yellow, and variably on femora I and II and tibiae I and II. *Orange*: Flagellum on basal half, scape, tibia III basally, 1st tergum apically, 2nd tergum except laterally, and 3rd tergum except apically and laterally. *Black*: Flagellum apical half, clypeus, mandible, coxae, 1st-trochanters blackish, variably on femora I and II and tibiae I and II, femur III, tibia III except basally, 1st abdominal tergum apically, 2nd tergum laterally blackish, 3rd tergum apically and laterally, and most of terga 4 to apex of abdomen. *Punctation* (Fig. 86:B-E): Face finely rugulose with dense, fine pits; frons densely pitted; temple with dense, indistinct pits; propleurum with dense, very fine, more or less distinct pits, surface very slightly rugulose; mesopleurum with mostly sparse, fine, more or less distinct pits, surface very slightly rugulose. *Shape*: Clypeal margin not upturned (Fig. 78:A); flagellum moderately stout, linear to apex, basal 3 segments short (Fig. 80:F,H); areola broadly hexagonal, area dentipara narrowed, apophysis sharp and distinct, propodeum mostly flat level dorsally, carinae weak; 1st tergum moderately stout (Fig. 84:B), postpetiole moderately widened, 1.5 times as wide as long; femur III moderately swollen (Fig. 79:D), ratio about 0.23; Rs more or less straight. *Other*: 3rd valvula moderately short, about as long as basal 3 hind tarsomeres; wings tinted.

REMARKS. See discussion under *monticola*.

RANGE. Alaska and Yukon Territory southward to Oregon, Idaho, and Colorado. Collections June through August.

MATERIAL EXAMINED. **Holotype**: male, Stone Mt. Pk., 3800+ ft., BRITISH COLUMBIA, H. and M. Townes (AEI). **Paratypes**: 54 M, 4 F (AEI, Dasch, INF, OSU): ALASKA, BRITISH COLUMBIA, COLORADO, IDAHO, OREGON, YUKON TERRITORY.

ETYMOLOGY. (Greek) *Penta-* (5) + *crocus* (orange), referring to the generally orange abdominal terga 2-6 of the male.

Endasys serratus Luhman, sp. n.
(Fig. 22:A-D; Map 20)

MALE DIAGNOSIS. Medium large to large, 6-7.5 mm long; 3 (or 4th partial) tyloids on flagellar segments 10-12 (or 13) long and thin, sharply angled anteriorly; clypeus black, margin more or less distinctly upturned (Fig. 78:C); propodeum mostly smooth between carinae, area dentipara distinctly narrowed mesally, widened laterally with acute angles, apophysis distinct; 1st abdominal tergum moderately slender (Fig. 84:C), postpetiole more or less elongate; mesopleurum mostly sparsely punctate with fine pits, surface shiny and smooth; punctation of head and thoracic pleura mostly fine to very fine, surfaces smooth and shiny; legs generally black, at least coxae, 1st-trochanters, and leg III; abdomen entirely orange on terga 2-6, black basally and apically, dark forms with increased black on terga 4-6; black scape, clypeus, and tegula; western United States and Canada

MALE DESCRIPTION. *Orange*: Trochantelli and abdomen on 1st tergum apically and terga 2-6; light forms with femora I and II and tibiae I and II yellowish orange , femur III and tibia III orange except apices black. *Black*: Flagellum, scape, clypeus, mandible, tegula, coxae, 1st-trochanters, leg III; dark forms usually with terga 4-6 at least apically black, legs I and II more blackish. *Punctation* (Fig. 86:A-D): Face very finely granular with very dense, fine pits, surface shiny; frons with moderately dense to moderately sparse punctation, fine pits; temple with moderately sparse, fine, indistinct pits; propleurum with mostly moderately sparse punctation, very fine, more or less distinct pits; mesopleurum with variably sparse, fine pits, surface smooth and shiny; postpetiole and 2nd tergum smooth and shiny. *Shape*: Clypeal margin more or less upturned (Fig. 78:C); flagellum moderately slender (Fig. 80:B), its segments mostly longer than wide to apex; areola broadly hexagonal, area dentipara distinctly narrowed mesally, widened laterally, apophysis distinct, propodeum mostly smooth between distinct carinae; 1st tergum moderately slender (Fig. 84:F), postpetiole appearing longer than wide, 2nd tergum regular; femur III moderately swollen (Fig. 79:C), ratio about 0.22. *Other*: 3 tyloids on flagellar segments 10-12, generally long and thin, anterior edge sharply angled, sometimes a partial tyloid on segment 13, and occasionally 2 tyloids on segments 10-11; wings hyaline to slightly tinted; 5th abdominal sternum complete apically (Fig. 81:B), or sometimes broken; glumes dense (Fig. 88:A).

FEMALE DIAGNOSIS. Medium large, 5-6 mm long; flagellum orange, distinctly narrowed apically (Fig. 80:E); clypeus black, impressed apically, margin more or less distinctly upturned (Fig. 78:C); head finely and unevenly punctate, propleurum with moderately dense, fine punctation, surface smooth and shiny; coxae blackish, femora blackish orange, abdomen weakly bicolored, apical half more or less blackish, remaining terga orange; 3rd valvula moderately long, as long as basal 4 hind tarsomeres; femur III swollen (Fig. 79:E), ratio about 0.29; 1st abdominal tergum moderately stout (Fig. 84:B), postpetiole moderately widened.

FEMALE DESCRIPTION. *Orange*: Flagellum, scape, mandible, tegula, legs I and II except coxae more blackish, coxa III apically, hind trochanters, femur III blackish orange, tibia III, hind tarsomeres, abdomen except apical 3 terga and laterally on terga 3-5. *Black*:

Clypeus, coxae I and II more orangish black basally, coxae III except apically, femur III more blackish orange, abdomen on apical 3 terga and laterally on terga 3-5. *Punctation* (Fig. 86:B,E): Face slightly rugulose with dense, uneven pits, surface shiny; frons with dense, variable pits; temple with moderately sparse, finely distinct pits; propleurum with moderately dense, finely distinct pits; mesopleurum with variably sparse, finely distinct pits, surface slightly rugulose. *Shape*: Clypeus slightly impressed apically, margin more or less distinctly upturned (Fig. 78:C); flagellum moderately slender, distinctly narrowed apically, basal 3 segments short (Fig. 80:E,H); areola elongate hexagonal anteriorly, area dentipara narrowed, apophysis distinct, propodeal carinae weak; 1st tergum stout (Fig. 84:A), postpetiole moderately widened, more than 1.5 times as wide as long, sides slightly diverging toward apex; femur III swollen (Fig. 79:E), ratio about 0.29. *Other*: 3rd valvula moderately long, about as long as basal 4 hind tarsomeres; wings tinted.

REMARKS. Males of this species differ from all others in the Monticola Group by the 3 or 4 angular tyloids and the distinctly upturned clypeal margin. It appears related to Palearctic *brevis* (Gravenhorst). Males of *serratus* differ from it by tergum 2 being smooth and shiny, not mat and punctate. Females differ by lacking a white band on the flagellum. In the Nearctic it may be most closely related to *monticola*. Female *serratus* differs from *monticola* by a longer 3rd valvula, about as long as the basal 4 hind tarsomeres, and the distinctly bicolored abdomen, weakly so in *monticola*. It differs from most other species in the group by the orange, distinctly narrowed flagellum.

RANGE. Pacific Northwest from British Columbia to northern California, eastward to Nevada and Idaho. Collections May through August.

MATERIAL EXAMINED. **Holotype**: male, Tuscarora, NEVADA, 5-vi-1978, H. and M. Townes (AEI). **Paratypes**: 96 M, 3 F (AEI, CNC, Dasch): BRITISH COLUMBIA, CALIFORNIA, IDAHO, NEVADA, OREGON, WASHINGTON. Other material studied: 6 M (CAS, OSU, UCR): CALIFORNIA, OREGON.

ETYMOLOGY. *Serratus* (saw-toothed), referring to the shape of the tyloids.

BICOLOR GROUP.

DIAGNOSIS. Clypeus distinctly wide, 3 or more times wider than high, margin more or less upturned, abdomen black or bicolored.

This is a small group of 7 species with characteristic widening of the face and clypeus, face width 1.2 times height of face plus clypeus. Other features of the group generally include clypeal margin sharp and more or less upturned, apophysis weak or absent, radial cell elongate, more or less curved apically, 5th abdominal sternum usually complete, abdomen with black at least apically, and anterior terga of males more or less mat. The range of this group is mostly boreal; some have been collected in northern Canada and in Greenland. One species, *nigrans*, occurs mostly in central and northern California. *E. minutulus* (Thomson) and *coriaceus* are Holarctic. Recorded hosts are tenthredinids.

This group corresponds to the Palearctic Eurycerus Group. Included in the Nearctic group are *bicolor* (Lundbeck), *callistus, coriaceus, declivis, euryops, minutulus* (Thomson), and *nigrans*.

Endasys bicolor (Lundbeck)
(Fig. 43:A-D; Map 7)

Phygadeuon bicolor Lundbeck, (1896) 1897: 227. **Lectotype** designated by Townes, 1961: 111.
Stylocryptus bicolor: Roman, 1916: 4; Hendricksen and Lundbeck, 1918: 525.
Endasys (Endasys) bicolor: Townes, 1944: 212; Townes and Townes, 1951: 246.
Endasys bicolor: Carlson, 1979: 417.

MALE DIAGNOSIS. Medium, 5.5-6 mm long; 3 short tyloids on flagellar segments 9-11; clypeus black, margin sharp but not upturned (Fig. 78:B); 1st abdominal tergum stout (Fig. 84:D), spiracle at midpoint, postpetiole longer than wide; 1st and 2nd terga mat; abdomen distinctly bicolored, black usually on terga 1, 3 apically, and 4 to apex; femora and tibiae mostly yellow-orange except apices of femur III and tibia III black, coxae and trochanters black; mesopleurum often sparsely punctate centrally; propodeum a little swollen dorsally, areola nearly hexagonal, apophysis weak; Greenland and northern North America.

MALE DESCRIPTION. *Yellowish orange*: Mandible mostly blackish orange, femora except femur III apically, and tibiae except tibia III apically. *Orange*: Anterior abdominal terga, usually 2nd tergum entirely, 3rd tergum basally, and sometimes 1st tergum apically. *Black*: Flagellum, scape, clypeus, tegula, coxae, 1st-trochanters, apices of femur III and tibia III, hind tarsomeres, abdomen basally and apical half, usually 3rd tergum apically, terga 4-apex. *Punctation* (Fig. 86:C,E): Face rugulose with very dense pits; frons a little rugulose with dense to very dense pits; temple with moderately sparse, indistinct pits; propleurum with mostly sparse, distinct pits, surface slightly rugulose; mesopleurum with sparse to very sparse punctation centrally, otherwise with mostly sparse, distinct pits, a little rugulose on lower part; 1st and 2nd terga slightly mat. *Shape*: Clypeal margin sharp but not upturned (Fig. 78:B); flagellum moderately stout (Fig. 80:C), its segments squarish on apical half; areola nearly or broadly hexagonal, area dentipara regular or a little narrowed, apophysis weak, propodeum more or less swollen dorsally, carinae distinct; 1st tergum stout (Fig. 84:D), spiracle at midpoint, postpetiole longer than wide, 2nd tergum regular; femur III moderately slender to moderately swollen, ratio 0.20-0.22; Rs straight (Fig. 82:B). *Other*: 3 short tyloids (Fig. 87:D) on flagellar segments 9-11; 5th abdominal sternum complete medially; and glumes dense (Fig. 88:A).

FEMALE DIAGNOSIS. Medium small, 5-6 mm long; face wide, more than 3 times as wide as long, rugulose with sparse to dense, coarse pits; clypeus black, margin distinctly upturned (Fig. 78:D); temple, propodeum, and femur III distinctly swollen; punctation of head and thoracic pleura generally with sparse, distinct pits, surfaces slightly rugulose; flagellum more or less bicolored dorsally, moderately stout, more or less linear to apex, basal 3 segments moniliform (Fig. 80:F-G); coxae and trochanters blackish, yellow-orange trochantelli, femora and tibiae; abdomen weakly bicolored, tergum 5 apically and terga 6-apex blackish, remaining terga orange; 3rd valvula moderately long, about as long as basal 4 hind tarsomeres; 1st tergum moderately stout (Fig. 84:B), postpetiole squarish; radial cell narrow, Rs straight (Fig. 82:B).

FEMALE DESCRIPTION. *Yellowish orange*: Flagellum ventral half, scape, mandible, femora and trochantelli, tibiae, and hind tarsomeres except 5th segment apically. *Orange*: Abdominal terga 1-4 and 5 basally. *Black*: Most of flagellum dorsally and apical half ventrally, clypeus, tegula, coxae, 1st-trochanters, and abdominal terga 5 apically and 6 to apex. *Punctation* (Fig. 86:D-E): Face rugulose with variably sparse to dense, coarse pits; frons with sparse to very sparse, coarse pits; temple with very sparse, more or less distinct pits; propleurum with variably dense to sparse punctation, more or less coarse pits, surface a little rugulose on lower part; mesopleurum with mostly sparse, distinct pits, surface slightly rugulose. *Shape*: Clypeal margin distinctly upturned (Fig. 78:D); face wide, more than 3 times as wide as high; temple swollen; flagellum short and moderately stout, more or less linear to apex, basal 3 segments nearly moniliform (Fig. 80:F-G); propodeum distinctly swollen dorsally, carinae weak, areola nearly hexagonal, area dentipara narrowed and high, apophysis weak, indistinct; 1st tergum moderately stout (Fig. 84:B), postpetiole moderately widened, nearly square; femur III swollen (Fig. 79:E), ratio about 0.31; radial cell narrow, Rs straight. *Other*: 3rd valvula moderately long, about as long as basal 4 hind tarsomeres; wings tinted.

REMARKS. See discussion under *coriaceus* and *melanurus*.

RANGE. Northern North America from Alaska and British Columbia eastward across the Northwest Territories and northern Quebec to Greenland. Collections June through August.

HOST. A couple of specimens were collected on "tundra" (CNC). They probably parasitize nematine species of *Pristiphora* sawflies on willows.

MATERIAL EXAMINED. **Lectotype**: female, Frederikshaab, Steenstrup, GREENLAND, 5-vi-1877 (ZMC) **Paralectotype**: female, GREENLAND, (ZMC). Other material studied: 16 M, 3F (AEI, CNC, ZMC): ALASKA, BRITISH COLUMBIA, GREENLAND, NORTHWEST TERRITORIES, QUEBEC.

ETYMOLOGY. *Bi-* (2) + *color*, probably referring to the bicolored abdomen.

Endasys callistus Luhman, sp. n.
(Fig. 34:A-D; Map 8)

MALE DIAGNOSIS. Medium large, 5-7 mm long; clypeus short and wide, margin not upturned; temple conspicuously swollen; abdomen bicolored, usually terga 4-apex of abdomen black; coxae I and III and trochanters I and II white ventrally, femur III and tibia III orange with black apices; punctation of head generally finely dense, frons with evenly dense, very fine but distinct pits; flagellum slender and a little elongate; eastern and northern United States

MALE DESCRIPTION. *White*: Scape, clypeus, mandible, tegula, coxae I and II except basally, trochanters except hind trochanters more orangish basally. *Yellow*: Front and middle coxae basally, femora I and II, tibiae I and II. *Orange*: Coxa III, femur III, tibia III except apices, usually terga 1-3 and basally on tergum 4. *Black*: Most of flagellum, femur III, tibia III apically, hind tarsomeres, abdomen on terga 4 apically and laterally, tergum 5 apically, terga 6-apex, often 1st tergum except apex. *Punctation* (Fig. 86:A-C): Face very finely granular or rugulose with very dense, fine pits; frons with evenly dense, finely distinct pits; temple with moderately dense, indistinct pits, surface more or less smooth and shiny; propleurum with dense to very sparse, fine pits; mesopleurum impunctate centrally, otherwise with variably dense punctation, surface a little rugulose; 1st and 2nd terga smooth and shiny. *Shape*: Clypeus short and wide, margin not upturned apically; flagellum slender (Fig. 80:A), all segments elongate; temple distinctly swollen; areola broadly hexagonal, area dentipara narrowed mesally, widened laterally (Fig. 83:C), apophysis distinct, propodeal carinae distinct; 1st tergum moderately slender (Fig. 84:F), postpetiole longer than wide, 2nd tergum wide; femur III moderately swollen (Fig. 79:C), ratio 0.22; radial cell long, Rs straight. *Other*: Wings hyaline, 5th abdominal sternum complete apically, and glumes variably dense.

FEMALE DIAGNOSIS. Medium large, about 6 mm long; flagellum short and slender, narrowed apically (Fig. 80:E), segments 5-9 white, remaining ones black; clypeus short and wide, mostly black, distinctly upturned (Fig. 78:D); temple distinctly swollen, punctation sparse with fine pits, frons evenly punctate with dense, very fine pits; abdomen bicolored, orange on terga 1-3 and 4 basally, black on 4 apically and 5-apex; 3rd valvula moderately long, as long as basal 4 hind tarsomeres; coxae I and II yellow, coxa III orange, femur III and tibia III orange with black apices; area dentipara narrowed, apophysis weak, propodeal carinae weak; eastern and northern United States

FEMALE DESCRIPTION. *White:* Flagellar segments 5-9. *Yellow:* Scape, mandible, tegula, and legs I and II. *Orange:* Clypeus with some blackish orange apically, leg III except femur and tibia apically and hind tarsomeres, abdominal terga 1-3 and 4 basally, sometimes tergum 3 apically. *Black:* Flagellar segments except annulus, femur III and tibia III apically, hind tarsomeres, tergum 4 apically and terga 5-apex, sometimes tergum 3 apically. *Punctation* (Fig. 86:B): Face finely rugulose with very dense pits; frons and propleurum with dense, fine pits; temple with sparse, fine, more or less distinct pits; mesopleurum variably punctate and variably rugulose. *Shape:* Clypeus short and wide, margin distinctly upturned (Fig. 78:D); flagellum short and slender, narrowed apically, basal 3 segments elongate (Fig. 80:E,I); temple distinctly swollen, face widened; areola broadly hexagonal, area dentipara narrowed, apophysis moderately distinct or weak, propodeal carinae weak; 1st tergum stout (Fig. 84:A), spiracle nearly at midpoint, postpetiole a little wider than long, nearly square; femur III swollen (Fig. 79:E), ratio about 0.29; radial cell long, Rs straight. *Other:* 3rd valvula moderately long, as long as basal 4 hind tarsomeres, wings tinted, and ocelli very small.

REMARKS. Male *callistus* is distinguished from its relative *declivis* by the white scape, clypeus, and coxae; females, by the basal 3 flagellar segments long, in addition to its eastern distribution.

RANGE. Eastern and northern half of the United States from Minnesota and Nebraska eastward to Maine and West Virginia. Collections May through July, and September and October.

HOST. Tenthredinidae: Nematinae: *Pikonema alaskensis* (Rohwer), yellowheaded spruce sawfly (UM).

MATERIAL EXAMINED. **Holotype:** male, Ann Arbor, MICHIGAN, 20-v-1962, H. and M. Townes (AEI). **Paratypes:** 22 M, 7 F (AEI, Dasch, MCZ, UM): MAINE, MICHIGAN, MINNESOTA, NEBRASKA, NEW YORK, OHIO, PENNSYLVANIA, WEST VIRGINIA.

ETYMOLOGY. (Greek) *Call-* (beautiful) + *-istus* (-est), referring to the very pretty combination of white, orange, and black.

Endasys coriaceus Luhman, sp. n.
(Fig. 42:A-D; Map 2)

MALE DIAGNOSIS. Medium small, 5-7 mm long; 2 long tyloids on flagellar segments 10 and 11, flagellum moderately slender (Fig. 80:B), segments a little longer than wide on apical half; clypeus black, margin distinctly upturned (Fig. 78:D); punctation of frons dense to sparse with indistinct pits; 1st and 2nd abdominal terga conspicuously and coarsely mat;

radial cell very long, Rs more or less straight; 1st tergum moderately slender to moderately stout, postpetiole longer than wide; abdomen bicolored, orange at most on 1st tergum apically, 2nd tergum entirely, and 3rd tergum basally; area dentipara regular or a little narrowed, apophysis more or less distinct; legs mostly yellow-orange with black coxae, 1st-trochanters, and apices of femur III and tibia III; Alaska, western Canada, and Sweden.

MALE DESCRIPTION. *Yellowish orange*: Mandible apically, femora and trochantelli except apices of femur III, tibiae except tiba III apically, often basally on hind tarsomeres, abdomen at most with 1st tergum apically, 2nd tergum entirely, and 3rd tergum basally. *Black*: Mandible basally, flagellum, scape, clypeus, tegula, coxae, 1st-trochanters, usually femur III and tibia III apically, most of hind tarsomeres except basally, and abdomen at least basally, tergum 3 apically, and terga 4-apex; dark forms with abdomen entirely black. *Punctation* (Fig. 86:B,D-F): Face rugulose with dense, coarse pits; frons with variably dense to sparse punctation, finely distinct pits; temple with sparse, fine, more or less distinct pits; propleurum variably punctate, often coarse and rugulose; mesopleurum very sparse to impunctate centrally, otherwise dense, coarse, and rugulose; 1st and 2nd terga usually distinctly mat, sometimes coarsely so. *Shape*: Clypeal margin distinctly upturned (Fig. 78:D); flagellum moderately slender (Fig. 80:B), segments a little longer than wide to apex; areola nearly hexagonal, area dentipara narrowed or regular, widened laterally, apophysis moderately distinct; 1st tergum moderately slender to moderately stout, postpetiole longer than wide, 2nd tergum regular or slightly narrowed; femur III moderately slender (Fig. 79:B), ratio 0.21-0.22; radial cell very long, Rs more or less straight. *Other*: Wings a little tinted, 5th abdominal sternum complete, glumes dense, 2 long tyloids (Fig. 87:A) on flagellar segments 10 and 11.

FEMALE DIAGNOSIS. Medium small, 5.5 mm long; 3rd valvula very long, about 1.25 times as long as hind tarsomeres; 1st and 2nd terga distinctly mat, abdomen mostly black with brown on 1st tergum apically and 2nd tergum basally; radial cell very long, Rs straight on about basal 0.75, distinctly curved on apical 0.25; flagellum mostly black, moderately slender and very short, slightly narrowed apically, basal 3 segments a little longer than wide; femur III moderately slender (Fig. 79:B), ratio 0.25; legs mostly brownish, coxae and 1st-trochanters black; face and clypeus wide, clypeal margin truncate, slightly upturned; face densely punctate with very fine pits; propodeum swollen and without apophysis.

FEMALE DESCRIPTION (based on one specimen). *Brownish*: Flagellum ventrally, scape, femora and trochantelli, tibiae, femur III and tibia III except darker apically, abdomen on 1st tergum apically and 2nd tergum basally. *Black*: Flagellum dorsally, clypeus, mandible, tegula, coxae, 1st-trochanters, most of abdomen except tergum 1 apically and tergum 2 basally. *Punctation* (Fig. 86:A-C): Face a little rugulose with dense, fine pits; frons with sparse to very sparse, finely distinct pits; temple with sparse, indistinct pits; propleurum with dense to sparse, very fine pits, mostly smooth and shiny; mesopleurum with mostly sparse, fine, indistinct pits, surface slightly rugulose; 1st and 2nd terga strongly mat. *Shape*: Clypeus wide, margin truncate and slightly upturned (Fig. 78:C); face wide, more than 3 times as wide as high; temple distinctly swollen; flagellum short and moderately slender, very slightly narrowed apically, basal 3 segments slightly longer than wide; propodeum distinctly swollen dorsally, carinae weak, areola broadly hexagonal, area dentipara narrowed and high, without apophysis; 1st tergum stout (Fig. 84:A), postpetiole square, spiracle at middle; femur III moderately slender (Fig. 79:C), ratio 0.25; radial cell very long, basal 0.75 of Rs straight, apical 0.25 distinctly curved. *Other*: 3rd valvula very long, 1.25 times longer than hind tarsomeres; wings tinted.

REMARKS. This species appears related to *bicolor*. Males differ from the latter by having 2 long tyloids, not 3; moderately slender flagellum; and abdominal terga 1-3 strongly mat. Females are distinguished by the very long 3rd valvula, 1.5 times longer than the hind tarsomeres.

RANGE. Alaska, British Columbia, Saskatchewan, and Sweden. Collections June through August.

MATERIAL EXAMINED. **Holotype**: male, Stone Mt. Pk., 3800 ft., BRITISH COLUMBIA, H. and M. Townes (AEI). **Paratypes**: 6 M, 1 F (AEI): ALASKA, BRITISH COLUMBIA, SASKATCHEWAN. Other material studied: 3 F (AEI): SWEDEN.

ETYMOLOGY. *Coriaceus* (leathery), referring to the appearance of abdominal terga 1-3.

Endasys declivis Luhman, sp. n.
(Fig. 45:A-D; Map 2)

MALE DIAGNOSIS. Large, about 8 mm long; clypeus black, short and wide, margin sharp and more or less upturned (Fig. 78:C); propodeum short and more or less sloped, area between carinae mostly smooth; 1st abdominal tergum stout and wide (Fig. 84:D), postpetiole square, 2nd tergum wide, terga 1 and 2 distinctly mat; legs orange except black coxae, 1st-trochanters, and tibia III apically; abdomen bicolored, black basally, 3rd tergum apically, and terga 4-apex, remaining terga orange; areola broadly hexagonal, barely twice as wide as long; mesopleurum very sparsely punctate on central area, otherwise variably punctate with distinct pits and some rugulosity; British Columbia.

MALE DESCRIPTION. *Orange*: Trochantelli, femora, tibia except tibia III apically, and basally on hind basitarsus. *Black*: Flagellum, clypeus, most of mandible, tegula, coxae, 1st-trochanters, sometimes femur III blackish, hind tarsomeres except basally on basitarsus, abdomen basally, 3rd tergum apically, and terga 4-apex. *Punctation* (Fig. 86:D-F): Face rugulose with very dense pits; frons a little rugulose with dense to very dense pits; temple with sparse, variably distinct pits; propleurum rugulose, with mostly dense, distinct pits; mesopleurum with very sparse punctation and slightly rugulose centrally, otherwise with variable, distinct pits, surface rugulose; 1st and 2nd terga distinctly mat. *Shape*: Clypeus short and wide, margin sharp, more or less upturned (Fig. 78:C); flagellum moderately stout (Fig. 80:C), its segments squarish on apical half; areola broadly hexagonal, not widened, about twice as wide as long, area dentipara distinctly narrowed, apophysis distinct, propodeum short, sloped apically, mostly smooth between distinct carinae; 1st tergum stout and wide (Fig. 84:D), postpetiole squarish, 2nd tergum widened; femur III moderately swollen (Fig. 79:C), ratio 0.22-.023; radial cell long and more or less narrow. *Other*: 2 unequal length tyloids on flagellar segments 10-11, wings more or less hyaline or tinted; 5th abdominal sternum complete medially; glumes dense.

FEMALE DIAGNOSIS. Moderately large, about 6.5 mm long; clypeus black, distinctly widened, margin more or less upturned (Fig. 78:C); face rugulose with variably dense, coarse pits; flagellum very short, slender, distinctly narrowed apically (Fig. 80:E); propodeum short, slightly sloped apically, area dentipara distinctly narrowed, elongate apico-laterally; legs orange with black coxae and 1st-trochanters; abdomen bicolored, black basally and terga 3 apically to apex.

FEMALE DESCRIPTION. *White*: Sometimes flagellar segments 6 or 7 to 9. *Orange*: Sometimes scape, legs except coxae and 1st-trochanters, and abdominal terga 1 apically to 3 basally. *Black*: Flagellum except for annulus, scape usually, clypeus, most of mandible except apex, tegula, coxae, 1st-trochanters, abdomen basally and terga 3 apically to apex. *Punctation* (Fig. 86:B,D-E): Face rugulose with variably dense, coarse pits, surface shiny; frons with variably sparse, distinct pits; temple with sparse, more or less distinct pits; propleurum with mostly dense, fine, more or less distinct pits, a little rugulose on lower part; mesopleurum with mostly sparse, fine, more or less distinct pits and some rugulosity; 1st tergum slightly mat. *Shape*: Clypeus distinctly widened, margin sharp and more or less upturned (Fig. 78:C); flagellum very short and slender, distinctly narrowed apically, basal 3 segments short (Fig. 80:E,H); temple distinctly swollen; areola broadly hexagonal, area dentipara distinctly narrowed, apophysis distinct, propodeum short and sloped apically; femur III swollen (Fig. 79:E), ratio 0.29-0.30; 1st tergum short and stout, postpetiole moderately wide, about 1.5 times as wide as long; radial cell long, and narrowed, Rs straight. *Other*: 3rd valvula moderately long, as long as basal 4 hind tarsomeres; wings more or less hyaline or slightly tinted.

REMARKS. See discussion under *callistus*.

RANGE. British Columbia. Collections in July.

MATERIAL EXAMINED. **Holotype**: male, Stone Mt. Pk., 3800+ ft., BRITISH COLUMBIA, 13-vii-1973, H. and M. Townes (AEI). **Paratypes**: 4 M, 4 F (AEI): BRITISH COLUMBIA. Other material studied: 3 M (AEI): BRITISH COLUMBIA.

ETYMOLOGY. *Declivis* (sloped downward), referring to the propodeum.

Endasys euryops Luhman, sp. n.
(Fig. 49:A-D; Map 6)

MALE DIAGNOSIS. Medium small, 4.5-6 mm long; Face wide, more than 2.5 times as wide as high, rugulose with dense pits; clypeus black, margin distinctly upturned (Fig. 78:D); abdomen mostly brownish with yellowish basally on terga 3-6; coxae and femur III mostly brownish; yellow scape, mandible, tegula, and trochanters; area dentipara narrowed, apophysis more or less distinct, propodeum short, carinae distinct; 1st tergum moderately stout (Fig. 84:E), postpetiole a little longer than wide; northern United States and southern Canada.

MALE DESCRIPTION. *Yellow*: Flagellum ventrally, scape, mandible, tegula, coxae apically, trochanters, femora I and II, tibiae I and II, femur III basally and apically, tibia III except apically, and abdomen on 1st tergum apically, 2nd variably, and 3rd to 6th basally. *Brownish*: Coxae except apically, femur III except apically and basally, tibia III apically, and most of abdomen except 1st tergum apically, 2nd variably, and 3rd to 6th basally. *Black*: Clypeus, flagellum dorsally. *Punctation* (Fig. 86:B-D): Face rugulose with dense pits; frons with moderately dense, fine, more or less distinct pits; temple with moderately sparse, indistinct pits; propleurum with dense to very sparse, very fine, more or less distinct pits; mesopleurum impunctate centrally, otherwise with variably distinct pits, surface with rugulosity; 1st and 2nd terga very slightly mat. *Shape*: Clypeal margin distinctly upturned (Fig. 78:D); face wide, more than 2.5 times as wide as high; flagellum moderately stout (Fig. 80:C), its segments squarish on apical half; areola widely to broadly hexagonal, area dentipara narrowed, apophysis more or less distinct, propodeum short with distinct carinae; 1st tergum moderately stout (Fig. 84:E), postpetiole square to a little longer than wide, 2nd tergum regular; femur III moderately swollen (Fig. 79:C), ratio 0.24-0.25. *Other*: Wings slightly tinted, 5th abdominal sternum membranous medially, and glumes sparse (Fig. 88:C).

FEMALE DIAGNOSIS. Small, 4.5-5.5 mm long; clypeus black, wide, margin distinctly upturned (Fig. 78:D); face wide, about 2.5 times as wide as long, punctation dense and fine, surface finely granular or finely rugulose, appearing shiny; coxae mostly brownish, trochanters yellow, abdomen mostly brown with yellow on terga 2 and 3; 3rd valvula moderately short, as long as basal 3 hind tarsomeres; femur III moderately swollen (Fig. 79:C), ratio 0.28; 1st tergum moderately stout (Fig. 84:B), postpetiole moderately widened, about 1.5 times as wide as long; flagellum mostly yellow, short and moderately stout, more or less linear to apex, basal 3 segments nearly moniliform (Fig. 80:F,G).

FEMALE DESCRIPTION. *Yellow*: Most of flagellum except dorsally on apical half, scape, mandible, tegula, trochantelli, femora I and II more blackish yellow, tibiae I and II, tibia III except apically, and abdomen on 1st tergum apically, 2nd entirely, and on 3rd to 6th basally. *Brownish*: Flagellum dorsally on apical half, coxae, 1st-trochanters, I and II except apically, femur III, tibia III apically, and most of abdomen except 1st tergum apically, 2nd entirely, and 3rd to 6th apically, and apex. *Black*: Clypeus. *Punctation* (Fig. 86:B,E): Face finely rugulose or finely granular with dense pits; frons with dense, finely distinct pits; temple with moderately fine, distinct pits; propleurum and mesopleurum with variably dense to sparse punctation, fine, more or less distinct pits, surfaces slightly rugulose. *Shape*: Face wide, 2.5 times as wide as long; clypeus wide, margin distinctly upturned (Fig. 78:D); flagellum short and moderately stout, more or less linear to apex, basal 3 segments nearly moniliform (Fig. 80:F-G); areola broadly hexagonal, area dentipara distinctly narrowed, apophysis distinct, propodeal carinae distinct; 1st tergum moderately stout (Fig. 84:B), postpetiole moderately wide, about 1.5 times as wide as long; femur III moderately swollen (Fig. 79:D), ratio 0.28-0.29; Rs slightly curved. *Other*: 3rd valvula moderately short, as long as basal 3 hind tarsomeres; wings tinted.

REMARKS. See discussion under *nigrans*.

RANGE. Northern United States and southern Canada from Alaska, British Columbia, Washington, and Oregon eastward through Minnesota and Michigan to Newfoundland and New Brunswick, southward to Maine and New York. Collections May through July.

HOST. Tenthredinidae: Nematinae: *Nematus ribesii* (Scopoli), imported currantworm (USNM from UCB).

MATERIAL EXAMINED. **Holotype**: male, Hovland, Forest Sta., Cook Co., MINNESOTA, 10-vii-1973 (UM). **Paratypes**: 17 M, 3 F (AEI, CNC, Dasch, UM, USNM): ALASKA, BRITISH COLUMBIA, MICHIGAN, MINNESOTA, NEW BRUNSWICK, NEWFOUNDLAND, NEW YORK, OREGON, WASHINGTON, YUKON TERRITORY. Other material studied: 17 F (AEI, MCZ, UCB, USNM): MAINE, MICHIGAN, MINNESOTA.

ETYMOLOGY. (Greek) *Eury-* (wide) + *ops* (face), referring to the wide face.

Endasys minutulus (Thomson)
(Fig. 50:A-D; Map 1)

Stylocryptus (Stylocryptus) minutulus Thomson, 1883: 872. **Lectotype** designated by
    Sawoniewicz (Luhman and Sawoniewicz, 1991).
Stylocryptus transverseareolatus nigripes Strobl, 1901: 149.
Stylocryptus (Endasys) fusciventris Habermehl, 1916: 377.
Endasys minutulus: Fitton, 1982; Luhman and Sawoniewicz, 1991.

MALE DIAGNOSIS. Small, 3.5-4 mm long; face rugulose with very dense, coarse pits; clypeus swollen, margin distinctly upturned (Fig. 78:D); abdomen black, coxae blackish, trochanters yellow, femur III black with yellow basally and apically, tibia III yellowish basally and black apically; propodeal carinae weak, areola broadly hexagonal, area dentipara narrowed mesally, widened laterally, apophysis indistinct; 1st tergum moderately slender (Fig. 84:F), postpetiole longer than wide; Alaska, British Columbia, and Europe.

MALE DESCRIPTION. *Yellow*: Scape and pedicel, first few flagellar segments ventrally, areas around tyloids, mandible, tegula more blackish yellow, trochanters, femora I and II at least basally and apically, femur III basally and apically, tibiae I and II, tibia III basally. *Black*: Most of flagellum, lighter ventrally; clypeus, coxae, femora I and II more yellowish black except basally and apically, femur III except basally and apically, tibia III apically, hind tarsomeres, abdomen. *Punctation* (Fig. 86:B-C): Face rugulose with very dense, coarse pits; frons with dense, finely distinct pits; temple with moderately sparse, indistinct pits; propleurum mostly impunctate or with very sparse, very fine pits marginally; mesopleurum impunctate centrally, otherwise mostly smooth and impunctate, surface a little rugulose on lower part; metapleurum rugulose; 1st and 2nd tergum slightly mat and slightly wrinkled. *Shape*: Clypeus swollen, margin distinctly upturned (Fig. 78:D); flagellum moderately stout (Fig. 80:C), its segments squarish on apical half; areola broadly hexagonal, area dentipara narrowed mesally, widened laterally; apophysis very weak, propodeal carinae weak; 1st tergum moderately slender (Fig. 84:F), postpetiole a little longer than wide, 2nd tergum regular or very slightly narrowed; femur III slender (Fig.

79:A), ratio 0.20; Rs slightly curved. *Other*: Tyloids (2) long (Fig. 87:A), not prominent, often yellowish near base; wings a little tinted; 5th abdominal sternum mostly membranous medially, sometimes complete apically (Fig. 81:B).

FEMALE DIAGNOSIS. Small, 3-3.5 mm long; face wide and rugulose with very dense, coarse pits; clypeus black, margin more or less upturned (Fig. 78:C); black abdomen, coxae, and femur III; yellowish trochanters, femora I and II and tibiae I and II; radial cell long, distinctly longer than 2nd-discoidal cell, Rs a little curved (Fig. 82:C); flagellum bicolored, apical half linear to apex (Fig. 80:F).

FEMALE DESCRIPTION. *Yellow*: Flagellum basal half, mandible, tegula, trochanters, femora I and II blackish yellow, tibiae except tibia III apically. *Black*: Flagellum apical half, clypeus, coxae, femur III, tibia III apically, hind tarsomeres, abdomen. *Punctation* (Fig. 86:B,E): Face a little rugulose with very dense, coarse pits; frons and temple with sparse, very fine, more or less distinct pits; propleurum and mesopleurum with mostly very sparse, very fine, more or less indistinct pits, surfaces slightly rugulose on lower part; metapleurum rugulose. *Shape*: Clypeal margin more or less upturned (Fig. 78:C); face wide, nearly 3 times as wide as high; flagellum short and stout, linear on apical half (Fig. 80:F), apical segments very slightly swollen, basal 3 segments short, often squarish; areola broadly hexagonal, area dentipara narrowed mesally, widened laterally, apophysis weak, propodeal carinae weak; 1st tergum moderately stout (Fig. 84:B), postpetiole gradually widened; femur III moderately swollen (Fig. 79:D), ratio 0.24-0.25; Rs curved, radial cell slightly widened (Fig. 82:D), distinctly longer than 2nd-discoidal cell. *Other*: 3rd valvula moderately long, as long as basal 4 hind tarsomeres; wings a little tinted.

REMARKS. This is a Holarctic species common in Europe. See discussion under *nigrans*.

RANGE. Alaska, British Columbia, and Europe. Collections late June through August.

MATERIAL EXAMINED. **Lectotype**: female, SWEDEN (UZIL); 11 M, 3 F (AEI, Dasch): ALASKA, BRITISH COLUMBIA. European material studied: 103 M, 20 F.

ETYMOLOGY. *Minut-* (small) + *-ulus* (diminutive suffix), referring to the small size.

Endasys nigrans Luhman, sp. n.
(Fig. 11:A-D; Map 10)

MALE DIAGNOSIS. Small , 4-5 mm long; mostly black with yellow scape, trochantelli, femora I and II anteriorly, and tibiae except tibia III apically; flagellum moderately stout, segments squarish on apical half (Fig. 80:C); face rugulose with very dense, coarse pits; black clypeus with margin more or less distinctly upturned (Fig. 78:C); propodeal carinae moderately strong, area dentipara narrowed mesally, widened laterally, apophysis moderately strong; 1st abdominal tergum moderately slender (fig. 84:F), postpetiole longer than wide; postpetiole and 2nd tergum smooth and shiny; 5th abdominal sternum membranous medially; western United States.

MALE DESCRIPTION. *Yellow*: Often scape, trochantelli, femur I anteriorly, often femur II anteriorly, tibiae except tibia III apically; light forms sometimes yellowish on terga 1 and

2 apically and clypeus. *Black*: Flagellum, usually scape more blackish, clypeus, mandible, tegula, coxae, 1st-trochanters, femora I and II mostly posteriorly, tibia III apically, hind tarsomeres, abdomen, but more brownish on terga 2-4. *Punctation* (Fig. 86:B,D-F): Face rugulose with very dense, coarse pits; frons and temple with dense, more or less distinct pits; propleurum sparsely punctate to impunctate, surface smooth and shiny; mesopleurum mostly impunctate or with very sparse punctation, otherwise with sparse, fine pits; postpetiole smooth and shiny, larger specimens with more distinct pits. *Shape*: Clypeal margin more or less distinctly upturned (Fig. 78:C); flagellum moderately stout (Fig. 80:C), segments square or a little longer than wide on apical half; areola broadly hexagonal, area dentipara narrowed mesally, widened laterally, apophysis moderately distinct; 1st abdominal tergum moderately slender (Fig. 84:F), postpetiole longer than wide, femur III moderately slender to moderately swollen, ratio 0.19-0.22; Rs more or less straight, curved at apex (Fig. 82:C). *Other*: tyloids (2) long (Fig. 87:A), not prominent; wings tinted; 5th abdominal sternum membranous medially; cranial suture visible below median ocellus; sometimes 1st and 2nd terga with fine wrinkles.

FEMALE DIAGNOSIS. Small, 4 mm long; clypeus usually black, margin more or less upturned (Fig. 78:C); face rugulose, pits dense and coarse; flagellum yellowish black, moderately stout (Fig. 80:F), more or less linear to apex; abdomen black basally and apically, more brownish on tergum 1 apically, tergum 2, and apically on terga 3-6; coxae and femur III blackish; yellow trochanters and tibia III except latter black apically; area dentipara narrowed, carinae weak, more or less distinct apophysis; Rs nearly straight; 1st tergum moderately, postpetiole moderately expanded, about 1.5 times as wide as long; 3rd valvula moderately long, about equal to basal 4 hind tarsomeres.

FEMALE DESCRIPTION. *Yellowish*: sometimes flagellum, scape, sometimes clypeus, mandible, tegula, sometimes coxae I and II apically, trochanters, I and II except latter posteriorly, tibiae except tibia III apically. *Brownish to Orangish*: 1st tergum apically, variably on terga 2 and 3, and basally on terga 4-6. *Black*: Flagellum more yellowish black, usually clypeus, coxae except I and II apically, femur II posteriorly, femur III, tibia III apically, hind tarsomeres, abdomen basally and apically, variably on terga 2 and 3, and apically on terga 4-6. *Punctation* (Fig. 86:B,D-E): Face rugulose with dense and coarse pits; frons with variably sparse to dense punctation, more or less distinct pits; temple with moderately sparse, distinct pits; propleurum with moderately dense, very fine pits; mesopleurum slightly rugulose with mostly sparse, more or less distinct pits. *Shape*: Clypeus slightly impressed apically, margin more or less upturned (Fig. 78:C); flagellum moderately stout, more or less linear to apex, only very slightly narrowed apically, basal 3 segments nearly moniliform (Fig. 80:F-G); areola broadly hexagonal, area dentipara narrowed, apophysis moderately distinct; 1st tergum moderately stout (Fig. 84:B), moderately expanded, postpetiole about 1.5 times as wide as long; femur swollen III (Fig. 79:E), ratio about 0.3. *Other*: 3rd valvula moderately long, about as long as basal 4 hind tarsomeres; wings very slightly tinted.

REMARKS. This species is closely related to *euryops* and *minutulus* (Thomson). Males are distinguished from *euryops* by the entirely black abdomen and tegula, never brownish; and from *minutulus* by being larger than 3 mm, and having the hind femur entirely black, not yellow basally and apically. Female *nigrans* is separated from *euryops* by the blackish yellow clypeus, and the coxae, hind femur, and abdomen mostly blackish, not brownish; and differs from *minutulus*, in addition to the distribution, by the blackish yellow clypeus and flagellum, basal 3 segments moniliform, and the abdomen not entirely black.

RANGE. California to Washington, eastward to Nevada and Colorado. Collections generally March through May in California, June through August elsewhere.

MATERIAL STUDIED. **Holotype**: male, Hyatt Reservoir, OREGON, 29-vi-1978, H. and M. Townes (AEI). **Paratypes**: 39 M, 4 F (AEI, Dasch, OSU, UCB, UCR): CALIFORNIA, COLORADO, OREGON, WASHINGTON. Other material studied: 8 M (AEI, CAS, Dasch, OSU, UCB, UCD): CALIFORNIA, NEVADA, OREGON, WASHINGTON.

ETYMOLOGY. *Nigrans* (dark-colored), referring to the mostly black coloration.

AURICULIFERUS GROUP

DIAGNOSIS. Clypeus sharp and upturned along entire margin, propodeum short and level, apophysis distinct or nearly absent, petiole slender and elongate, postpetiole square or elongate, dorsomedian carina weak or nearly absent, face slightly widened—about 1.1 times wider than height of face and clypeus, punctation of face evenly dense to very dense.

This is a small group of 7 species; *arkansensis* is doubtfully included. The group appears to be the northern counterpart of the Santacruzensis Group. Several of the species resemble species of *Amphibulus*. The most unifying characters of the group are the following: sexual dimorphism and dichromatism less pronounced, sexes easily associated, clypeal margin sharply upturned (Fig. 78:D); face a little widened, width about 1.1 times height of face and clypeus; punctation of face characteristically evenly dense, pits coarse, surface appearing granular or rugulose; male mesopleural plate mostly smooth and impunctate; propodeum short, but not sloped posteriorly, apophysis often toothlike, but may be reduced; 1st abdominal tergum slender, petiole and postpetiole elongate, dorsomedian carina reduced; hind femur of male mostly moderately slender, that of female swollen; 5th abdominal sternum often membranous medially, but never 6th.

Species occur throughout much of the U.S. and southern Canada, but not in the Southwest or Mexico. *E. arkansensis* has been reared from diprionid species; there are no other host records.

This group corresponds to the Palearctic Rusticus Group. Nearctic species included are *arkansensis*, *aurantifex*, *auriculiferus* (Viereck), *elegantulus*, *granulifacies*, *rugiceps*, and *spicus*.

Endasys arkansensis Luhman, sp. n.
(Fig. 70:A-D; Map 1)

MALE DIAGNOSIS. Large, 7-8 mm long; face very finely granular or very finely rugulose with very dense pits; clypeus yellow-orange, margin sharp and distinctly upturned (Fig. 78:D); flagellum long and slender (Fig. 80:A), all segments elongate; area dentipara slightly narrowed, apophysis strong and sharp, propodeum very slightly swollen dorsally, carinae moderately distinct; abdomen mostly orange, sometimes blackish on 2nd tergum basally, and 3rd tergum apico-laterally; leg III orange with white trochanters and orangish black hind tarsomeres; white coxae I and II, trochanters I and II, tibiae I and II dorsally; eastern United States

MALE DESCRIPTION. *White*: Scape, tegula, coxae I and II except basally, trochanters, tibiae I and II dorsally. *Yellow*: coxae I and II basally, femora I and II, tibiae I and II except latter dorsally. *Orange*: Clypeus and mandible more yellow-orange, flagellum pale orange ventrally, coxae III, femur III, tibia III except sometimes apically, abdomen except often most of 1st tergum, sometimes 2nd basally and 3rd apico-laterally. *Black*: Flagellum dorsally, sometimes hind tibia apically, hind tarsomeres more orangish black, 1st tergum often except apically, and sometimes 2nd basally and 3rd apico-laterally. *Punctation* (Fig. 86:A-D): Face finely granular or finely rugulose with very dense, fine pits; frons with dense, finely distinct pits; temple with mostly moderately dense, indistinct pits; propleurum with mostly sparse punctation, very fine, more or less distinct pits, rugulose on lower part; mesopleurum very sparsely punctate centrally with véry fine pits, otherwise variably punctate with more or less distinct pits, surface a little rugulose on lower part; 1st and 2nd terga smooth and shiny. *Shape*: Clypeal margin sharp, more or less distinctly upturned (Fig. 78:C); flagellum long and slender (Fig. 80:A), all segments elongate; areola broadly hexagonal, area dentipara slightly narrowed, apophysis strong and sharp, propodeal carinae distinct; 1st tergum moderately stout to moderately slender, postpetiole squarish, slightly swollen, dorsomedian carina more or less indistinct, 2nd tergum widened; femur III moderately swollen (Fig. 79:C), ratio 0.22-0.23; radial cell straight, Rs very slightly bowed. *Other*: Wings more or less hyaline, 5th abdominal sternum membranous medially, and glumes dense.

FEMALE DIAGNOSIS. Large, 7-8 mm long; clypeus orange, margin sharp and distinctly upturned (Fig. 78:D); cheek a little swollen; flagellum moderately stout, more or less linear to apex, basal 3 segments short (Fig. 80:F,H); propodeum and postpetiole a little swollen, areola broadly hexagonal, about twice as wide as long, area dentipara sloped, apophysis more or less distinct; 1st tergum stout and wide, postpetiole more than twice as wide as long (Fig. 84:A); 3rd valvula short, as long as basal 2 hind tarsomeres; femur III swollen (Fig. 79:E), ratio 0.30; leg III mostly orange with blackish on tibia apically and tarsomeres.

FEMALE DESCRIPTION. *Yellow*: Flagellar segments 1-9, 5-9 often paler, scape, tegula, legs I and II. *Orange*: Clypeus apically, mandible, leg III except often tibia apically and hind tarsomeres, abdomen except basally. *Black*: Flagellum apical half, sometimes more orangish black, tibia III apically, hind tarsomeres, 1st tergum except apically. *Punctation* (Fig. 86:E): Face and frons rugulose with very dense, coarse pits; temple with variably sparse to dense, distinct pits; propleurum and mesopleurum rugulose with dense pits. *Shape*: Clypeal margin sharp and upturned (Fig. 78:D); flagellum slightly clavate, more or less linear to apex, or very slightly narrowed, basal 3 segments short (Fig. 80:F,H); areola broadly hexagonal, twice as wide as long, area dentipara distinctly narrowed, apophysis distinct, propodeum swollen dorsally and a little sloped apically, carinae moderately weak; 1st tergum stout (Fig. 84:A), postpetiole slightly swollen and distinctly widened, more than twice as wide as long; femur III swollen (Fig. 79:E), ratio 0.30; Rs mostly straight, very slightly bowed (Fig. 82:A). *Other*: 3rd valvula short, as long as basal 2 hind tarsomeres; wings more or less hyaline or slightly tinted.

REMARKS. This species is not clearly related to other species in the Aurculiferus Group, differing especially by the stouter 1st abdominal tergum, widened 2nd tergum, slightly swollen postpetiole dorsally, and females with a very short 3rd valvula, about as long as the basal 2 hind tarsomeres. It is included here because of the sharply upturned clypeal margin and the sharply projecting apophysis. It may belong in the Subclavatus Group where it is similar to *patulus*. Males differ by the slightly swollen postpetiole, whiter coxae, and occasional black on abdominal tergum 2. Females differ from *patulus* by the flagellum nearly linear to the apex, basal 3 segments short, the slightly swollen propodeum and postpetiole, and the sharply projecting apophysis.

RANGE. Arkansas, Ohio, Tennessee, West Virginia, and Maryland. Collections April through July.

HOST. Diprionidae: *Neodiprion* sp. (USNM); *N. taedae linearis* Ross, loblolly pine sawfly (UCR from UAM).

MATERIAL EXAMINED. **Holotype**: male, White House, TENNESSEE, 24-25-v- 1981, B. and C. Dasch (AEI). **Paratypes**: 21 M, 20 F (AEI, Dasch, UCD, USNM): ARKANSAS, KENTUCKY, LOUISIANA, MARYLAND, OHIO. Other material studied: 3 M (USNM): ARKANSAS.

ETYMOLOGY. From Arkansas, the locality of much of the paratype series.

Endasys aurantifex Luhman, sp. n.
(Fig. 33:A-D; Map 12)

MALE DIAGNOSIS. Medium size, 6 mm long; light forms with orange legs, and abdomen except basally and terga 4-7 apically; dark forms with mostly black coxae, trochanters, abdomen except variably on terga 2 and 3, and apically on femur III and tibia III; slender legs and abdomen; clypeal margin sharply upturned (Fig. 78:E); face rugulose with evenly dense, coarse pits; propodeal carinae weak, area dentipara wide, apophysis weak but distinct; 1st abdominal tergum slender (Fig. 84:G), postpetiole distinctly elongate, 2nd tergum narrowed or regular; western half of northern United States and southern Canada.

MALE DESCRIPTION. *Orange* (light forms and type): Scape, clypeus, mandible, legs, tegula, and abdomen except basally, apically, and on terga 4-6 apically; dark forms only with orange scape, mandible apically, and legs except coxae and apices of femur III and tibia III black; clypeus, coxae, tegula, and terga 4-6 vary from orange to black. *Black* (light forms): Flagellum, abdomen basally, apically, and sometimes blackish on terga 4-6 apically; dark forms with black clypeus, mandible, coxae and trochanters, femur III and tibia III apically, and most of abdomen except variably on terga 2 and 3; sometimes most of femur III, entire abdomen black. *Punctation*: Face variably rugulose with evenly dense, coarse pits, surface shiny; frons variably sparse to dense with fine pits, usually rows horizontal and sparsely spaced, pits densely spaced; temple moderately sparse with indistinct pits; propleurum variably dense to sparse, or nearly impunctate inferiorly; mesopleurum impunctate centrally, otherwise with variably fine, more or less distinct pits; 1st and 2nd terga smooth and shiny. *Shape*: Flagellum moderately slender, segments longer than wide to apex; clypeal margin sharply upturned; areola broadly hexagonal, elongate apically, area dentipara wide, apophysis weak but distinct, propodeum short, 1st tergum slender, postpetiole distinctly elongate, 2nd tergum narrowed or regular; femur III moderately slender, ratio 0.22-0.23. *Other*: Wings tinted, 5th abdominal sternum complete medially, hind tibial spines short.

FEMALE DIAGNOSIS. Medium small, 5-6 mm long; mostly orange abdomen with apex blackish orange; clypeal margin sharply upturned; 3rd valvula very short, about as long as basal 2 hind tarsomeres; flagellum orange, moderately slender, narrowed apically (Fig. 80:E); face rugulose, frons and temple sparsely punctate; 1st abdominal tergum moderately slender (Fig. 84:C); femur III swollen (Fig. 79:E), ratio about 0.30.

FEMALE DESCRIPTION. *Orange* (light forms): Flagellum, scape, mandible, tegula, legs, and abdomen except often terga 4-apex blackish orange; dark forms with coxae, femur III, and apical terga more blackish. *Black* (light forms): Clypeus and terga 4-apex more orangish black; dark forms with blackish coxae, 1st-trochanters, femur III, tibia III apically, and terga 2-4 more brownish. *Punctation*: Face rugulose with coarse pits, surface shiny; frons variably sparse to dense with fine pits; temple sparse with more or less fine pits; propleurum moderately dense to moderately sparse with fine pits; mesopleurum nearly impunctate centrally and slightly rugulose, otherwise sparse with fine pits. *Shape*: Flagellum moderately slender, narrowed apically, basal 3 segments nearly moniliform; clypeal margin sharply upturned; areola elongate hexagonal, area dentipara narrowed, apophysis moderately distinct, propodeal carinae weak, propodeum very slightly swollen dorsally; femur III swollen, ratio about 0.30; radial cell short, Rs curved. *Other*: 3rd valvula short, about as long as basal 2 hind tarsomeres; wings tinted.

REMARKS. This species is related to *granulifacies* and *rugiceps*. Male *aurantifex* is distinguished from male *granulifacies* by abdominal terga 3-6 partly or entirely black, and by the moderately slender flagellum with long tyloids; and from *rugiceps* males by the orange or black coxae, black on terga 3-6, and the orange scape, not yellow. Female *aurantifex* is separated from both species by the orange flagellum with basal 3 segments moniliform, the mostly orange coxae I and II, and the curved Rs.

RANGE. Western half of United States and southern Canada from Alberta to Oregon and Minnesota, southward through Idaho and Montana to Nebraska and Colorado. Collections May through August.

MATERIAL EXAMINED. **Holotype**: male, Selma, OREGON, 13-v-1978, H. and M. Townes (AEI). **Paratypes**: 78 M, 27 F (AEI, CAS, CNC, UCB): ALBERTA, CALIFORNIA, COLORADO, IDAHO, MINNESOTA, MONTANA, NEBRASKA, OREGON. Other material examined: 6 M, 1 F (AEI, OSU, USNM): COLORADO, OREGON.

ETYMOLOGY. *Auranti-* (orange) + *fex* (maker), referring to orange color of the light forms.

Endasys auriculiferus (Viereck)
(Fig. 44:A-D; Map 3)

Phygadeuon (Bachia?) auriculiferus Viereck, (1916) 1917: 336. **Holotype** by monotypy, here determined.

Endasys (Endasys) auriculiferus: Townes, 1944: 212; Townes and Townes, 1951: 246.

Endasys auriculiferus: Carlson, 1979: 417.

MALE DIAGNOSIS. Medium large, 5.5-7 mm long; 3 equal-sized tyloids on flagellar segments 9-11; clypeal margin sharp and upturned (Fig. 78:D), clypeus black and orangish; 1st abdominal tergum slender (Fig. 84:G) and shiny, postpetiole square to elongate, area dentipara distinctly narrowed, apophysis distinct; whitish scape, coxae, and trochanters; abdomen mostly orange; leg III mostly orange except whitish on coxa and trochanter and black on tibia, at least apically; eastern United States and southern Canada.

MALE DESCRIPTION. *White*: Scape, coxae I and II, coxa III at least apico-ventrally, trochanters, tibiae I and II dorsally. *Yellow*: Legs I and II except whitish areas. *Orange*: Most of flagellum ventrally, apical half of clypeus, mandible, tegula, coxa III except apically, femur III, tibia III except apically, abdomen except apically, and sometimes basally. *Black*: Basal half of clypeus, tibia III apically, abdomen apically, and often basally; darker forms with more black on clypeus and leg III. *Punctation* (Fig. 86:A-D): Face dense and rugulose; frons with dense, more or less distinct pits, temple with dense, indistinct pits; prothorax densely punctate to impunctate, pits fine and indistinct, mesopleurum impunctate and shiny centrally, elsewhere with sparse, more or less distinct pits, slightly rugulose below; abdominal terga 1 and 2 shiny, pits sparse and inconspicuous. *Shape*: Clypeus sharp and upturned (Fig. 78:D), flagellum moderately stout (Fig. 80:C), its segments nearly square on apical half, areola widely hexagonal, area dentipara distinctly narrowed with strong carinae, apophysis sharp and distinct, 1st abdominal tergum slender (Fig. 84:G), postpetiole square to elongate, 2nd tergum nearly regular, femur III moderately swollen (Fig. 79:C), ratio about 0.23, Rs straight. *Other*: 3 moderately short, prominent tyloids (Fig. 87:C) on flagellar segments 9-11; wings hyaline to slightly tinted; and 5th abdominal sternum membranous medially.

FEMALE DIAGNOSIS. Medium large, 5.5-7 mm long; flagellum more or less linear to apex (Fig. 80:F), pale yellowish on segments 40 or 5-10; clypeus black, margin distinctly upturned (Fig. 78:D); 1st abdominal tergum slender (Fig. 84:C), postpetiole gradually widened, nearly square; face rugulose with very dense, coarse pits; propleurum with evenly dense, finely distinct pits; 3rd valvula long, as long as basal 4 hind tarsomeres; orange scape, clypeus sometimes, leg III except trochanter yellow, and abdomen; legs I and II and hind trochanters yellowish.

FEMALE DESCRIPTION. *Yellow*: Flagellar segments 4 or 5 to 9 pale yellow, legs I and II, trochanter III, often hind tarsomeres. *Orange*: Basal 3 or 4 flagellar segments, clypeus sometimes, scape, mandible, tegula, coxa III, femur III except sometimes more blackish orange, tibia III except apically, often hind tarsomeres. *Black*: Usually clypeus, femur III sometimes more blackish, tibia III except apically, sometimes hind tarsomeres. *Punctation*

(Fig. 86:B,D,F): Face rugulose with dense, coarse pits; frons less rugulose with very dense, coarse pits; temple with dense, more or less distinct pits; propleurum with evenly dense, fine pits. *Shape*: clypeal margin sharp and distinctly upturned (Fig. 78:E); short; areola a little elongate-hexagonal, often appearing triangular with truncate corners, area dentipara distinctly narrowed, vertically projecting apophysis sharp and distinct, propodeal carinae distinct; 1st tergum slender (Fig. 84:C), postpetiole square or less than 1.5 times as wide as long; femur III moderately swollen (Fig. 79:D), ratio 0.25; radial cell distinctly longer than 2nd-discoidal cell, Rs mostly straight. *Other*: 3rd valvula long, about as long as hind tarsomeres; wings tinted.

REMARKS. A black form occasionally occurs in which the orange and yellow colored areas are black, and the following areas white: clypeus, coxae apically, trochanters, tibiae I and II dorsally, and tarsomeres except apical segments. Morphologically it is indistinguishable from the usual orange forms. No female matching the black form has been found, and males are only from sites where the orange form was extensively collected. Specimens of the dark form are only from North Carolina, Virginia, Maryland, Ohio, and New York. Both black and orange forms of *E. auriculiferus* are easily distinguished from most species by a 3rd tyloid on segment 9, and the sharply upturned clypeal margin.

Other species with a 3rd tyloid on segment 9 are mostly black and orange with a bicolored abdomen. Its closest Palearctic relative appears to be *E. nitidus* (Habermehl), similar to the black form. Its closest Nearctic relative may be *spicus*. Females differ from *nitidus* by a combination of the tricolored (yellow, pale yellow, black), linear flagellum, narrow area dentipara with strongly projecting apophysis, and 3rd valvula about as long as basal 4 hind tarsomeres; and differs from female *spicus* by lacking distinctly white flagellar segments 5-9, and 3rd valvula shorter than hind tarsus.

RANGE. Eastern half of U.S. and southern Canada. Collections May to August and into September.

MATERIAL EXAMINED. **Holotype**: male, Putnam, CONNECTICUT, 12-vii-1905, H.L. Viereck (USNM). No paratypes. Other material studied: 1041 M, 304 F (AEI, CNC, Dasch, FLDA, OSU, UCB, UCD, UM, USNM): ARKANSAS, CONNECTICUT, FLORIDA, IOWA, KANSAS, KENTUCKY, MAINE, MARYLAND, MASSACHUSETTS, MICHIGAN, MINNESOTA, MISSOURI, NEBRASKA, NEW BRUNSWICK, NEW HAMPSHIRE, NEW JERSEY, NEW YORK, NORTH CAROLINA, OHIO, ONTARIO, PENNSYLVANIA, PRINCE EDWARD ISLAND, QUEBEC, RHODE ISLAND, SOUTH CAROLINA, SOUTH DAKOTA, TENNESSEE, VERMONT, VIRGINIA, WASHINGTON D.C., WEST VIRGINIA.

ETYMOLOGY. *Auriculi-* (little ear) + *ferus* (bearer), referring to the vertically projecting apophysis resembling upturned ears.

Endasys elegantulus Luhman, sp. n.
(Fig. 36:A-D; Map 5)

MALE DIAGNOSIS. Small, 3:5-4 mm long; mostly black abdomen, 1st tergum slender (Fig.84:G), white coxae and trochanters, femur III and tiba III yellow with black dorso-apically, white scape and clypeus, frons a little swollen, punctation of frons finely dense, area dentipara regular, apophysis weak and inconspicuous; southern Canada and eastern United States.

MALE DESCRIPTION. *White*: Scape, clypeus, often face just above clypeus, mandible, tegula, coxae except coxa III basally, trochanters, sometimes femora I and II and tibiae I and II anteriorly, sometimes tibiae I and II dorsally. *Yellow*: Flagellum ventrally, usually most of femora I and II, tibiae I and II except often dorsally and anteriorly, coxa III basally, femur III and tibia III except dorso-apically, hind tarsomeres often mostly yellowish, abdominal terga 1-3 on apical margin, and terga 4 and 5 basally. *Black*: Flagellum dorsally, coxa III dorso-basally, femur III and tibia dorso-apically, abdomen except terga 1-3 on apical margin, and terga 4 and 5 basally, and sometimes hind tarsomeres mostly blackish. *Punctation* (Fig. 86:A-B,D-E): Face finely granular; frons with variably dense to sparse, very fine pits, surface slightly swollen; temple with moderately dense, more or less indistinct pits; propleurum with dense and very fine punctation at margins, mostly impunctate, smooth and shiny on lower part; mesopleurum impunctate and slightly rugulose, otherwise with sparse, distinct pits, surface a little rugulose; 1st and 2nd terga smooth and shiny. *Shape*: Clypeal margin sharp, margin slightly upturned (Fig. 78:C); flagellum moderately stout (Fig.80:C), its segments squarish on apical half; areola broadly hexagonal, area dentipara regular, apophysis weak and indistinct, propodeal carinae distinct; 1st tergum slender (Fig. 84:G), nearly parallel- sided, postpetiole distinctly elongate, 2nd tergum regular; femur III moderately swollen (Fig. 79:C), ratio 0.23-0.25; Rs very slightly curved. *Other*: Tyloids (2) elongate (Fig. 87:A) and not prominent, wings hyaline, 5th abdominal sternum membranous medially, glumes sparse.

FEMALE DIAGNOSIS. Small, 3.5-4 mm long; abdomen shiny black, 1st tergum moderately slender (Fig. 84:C), clypeal margin sharp and distinctly upturned (Fig. 78:C), face rugulose with coarse punctation, punctation of frons, temple, and thoracic pleura mostly sparse with finely distinct pits.

FEMALE DESCRIPTION. *Yellow*: Flagellum basal half, scape, clypeus, mandible, legs except leg III more blackish yellow on coxa and femur, abdomen often blackish yellow on terga 4-apex. *Black*: Flagellum apical half, coxa III basally, femur III dorsally blackish yellow, and most of abdomen except often apically more blackish yellow. *Punctation* (Fig. 86:B,F): Face rugulose with dense, coarse pits; frons with variably dense to sparse, fine pits, surface a little swollen; temple with sparse, fine pits; propleurum with mostly sparse, fine pits; mesopleurum with mostly sparse, fine pits, surface slightly rugulose on lower part. *Shape*: Clypeal margin sharp and distinctly upturned (Fig. 78:C); flagellum moderately stout, linear to apex, basal 3 segments short (Fig. 80:F,H); areola nearly hexagonal, area dentipara narrowed, apophysis weak, propodeal carinae weak; 1st tergum moderately slender, gradually widened, postpetiole squarish or a little wider than long; femur III moderately swollen (Fig. 79:D), ratio 0.25-0.27; Rs slightly curved. *Other*: 3rd valvula moderately long, as long as basal 4 hind tarsomeres; wings hyaline; abdominal terga highly polished.

REMARKS. This species differs from all other species in the Auriculiferus Group by the small size, the mostly black abdomen, and the (males) white or (females) pale yellow coxae.

RANGE. British Columbia to Ontario and Michigan, southward to Kansas, Ohio, and South Carolina. Collections May and June.

MATERIAL EXAMINED. **Holotype**: male. Crystal Falls, MICHIGAN, 25-vi-1969, H. and M. Townes (AEI). **Paratypes**: 55 M, 9 F (AEI, Dasch, USNM): BRITISH COLUMBIA, KANSAS, MICHIGAN, OHIO, SOUTH CAROLINA. Other material studied: 3 M, 1 F (FLDA, MCZ): FLORIDA, MASSACHUSETTS.

ETYMOLOGY. *Elegant-* (elegant) + *-ulus* (diminutive suffix), referring to the small size and shiny black coloration with white.

Endasys granulifacies Luhman sp. n.
(Fig. 3:A-D; Map 7)

MALE DIAGNOSIS. Medium, 6-7 mm long; orange abdomen and legs beyond 1st-trochanters; black clypeus, mandible, often scape, 1st-trochanters, and hind tarsomeres; clypeal margin sharply upturned (Fig. 78:E); frons appearing granular with very dense, fine pits; propodeal carinae moderately strong, apophysis weak; Pacific Northwest.

MALE DESCRIPTION. *Orange*: Sometimes scape, legs usually beyond 1st-trochanters except hind tarsomeres, and abdomen except basally and apically. *Black*: Flagellum, usually scape, clypeus, mandible, tegula, coxae, usually 1st-trochanters, most of hind tarsomeres, and abdomen basally and apically. *Punctation* (Fig. 86:B,D-E): Face very dense with fine pits, surface appearing granular; frons densely pitted; temple with variable, more or less distinct pits; propleurum densely punctate to nearly impunctate, distinctly pitted, surface with rugulosity; mesopleurum impunctate centrally, otherwise with mostly sparse, fine, more or less distinct pits, surface usually rugulose; postpetiole and 2nd tergum sometimes wrinkled. *Shape*: Clypeus with margin sharply upturned (Fig. 78:E); flagellum moderately stout (Fig. 80:C), its segments square to a little longer than wide on apical half; areola broadly hexagonal, but not very wide; area dentipara narrowed and elongate apically, apophysis weak, propodeum slightly sloped; 1st abdominal tergum moderately slender (Fig. 84:F), postpetiole longer than wide; 2nd tergum regular; femur III moderately swollen (Fig. 79:C), ratio about 0.24. *Other*: Wings a little darkened; 5th abdominal sternum complete; sometimes very small 3rd tyloid on 12th flagellar segment.

FEMALE DIAGNOSIS. Medium, 6-7 mm long; orange scape, tegula, abdomen, and legs except blackish coxae, face rugulose with very dense, coarse pits; clypeal margin distinctly upturned (Fig. 78:D); flagellum bicolored, slender, and narrowed apically, first 3 segments short; femur III swollen (Fig. 79:E), ratio 0.29.

FEMALE DESCRIPTION. *Orange*: Flagellum on basal half, scape, mandible, legs except most of coxae, and abdomen. *Black*: Flagellum apical half, clypeus, coxae blackish. *Punctation* (Fig. 86:D-E): Face and frons rugulose with very dense, coarse pits; temple with sparse, coarse pits; propleurum with dense, distinct, but finer pits; mesopleurum rugulose with variably distinct pits. *Shape*: Clypeal margin distinctly upturned (Fig. 78:D); flagellum short, narrowed apically, first 3 segments short (Fig. 80:E,H); areola broadly hexagonal; area dentipara narrowed, apophysis weak to moderately distinct; 1st abdominal tergum moderately stout (Fig. 84:B), postpetiole moderately expanded, about 1.7 times as wide and long; femur III swollen (Fig. 79:E), ratio 0.29; radial cell long, Rs straight. *Other*: 3rd valvula moderately short, about as long as basal 3 hind tarsomeres.

REMARKS. See discussion under *aurantifex*.

RANGE. Pacific Northwest. Collections April through July.

MATERIAL EXAMINED. **Holotype**: male, Lake Tahoe, CALIFORNIA, 19-vi-1936, R.M. Bohart (AEI). **Paratypes**: 24 M, 1 F (AEI, OSU, UCB, UCD): CALIFORNIA, OREGON, WASHINGTON. Other material studied: 5 M (CAS): CALIFORNIA.

ETYMOLOGY. *Granuli-* (little grains) +*facies* (face), referring to the granular face.

Endasys rugiceps Luhman, sp. n.
(Fig. 15:A-D; Map 15)

MALE DIAGNOSIS. Medium large, 6-7.5 mm long; head rugulose with very dense, coarse pits; clypeus black, margin sharp and more or less upturned (Fig. 78:C); area dentipara narrowed, apophysis weak; 1st abdominal tergum slender (Fig. 84:G); white coxae I and II, coxa III ventrally, and trochanters; orange abdomen, femur III and tibia III, hind tarsomeres blackish orange; eastern half of United States and southern Canada.

MALE DESCRIPTION. *White*:.Sometimes clypeus, tegula, coxae I and II except basally, coxa III ventrally, trochanters. *Yellow*: Usually clypeus, coxae I and II basally, coxa III except ventrally, femora I and II, tibiae I and II. *Orange*: Flagellum more basally, sometimes clypeus blackish orange apically, usually mandible, femur III, tibia III, hind tarsomeres more blackish orange, and abdomen basally. *Black*: Flagellum more apically, usually clypeus, sometimes mandible, abdomen basally. *Punctation* (Fig. 86:E-F): Head rugulose with very dense, coarse pits; propleurum densely punctate to impuntcate, distinct pits, some rugulosity on lower half; mesopleurum impunctate centrally with some rugulosity, otherwise mostly densely pitted; 1st and 2nd terga mostly smooth and shiny. *Shape*: Clypeal sharp and more or less distinctly upturned along entire margin (Fig. 78:C); flagellum moderately slender (Fig. 80:B), its segments longer than wide to apex; areola broadly hexagonal, area dentipara narrowed, apophysis weak, propodeal carinae moderately distinct; 1st tergum slender (Fig. 84:G), postpetiole longer than wide or squarish, 2nd tergum regular; femur III moderately swollen (Fig. 79:C), ratio 0.22-0.24; radial cell widened, Rs straight. *Other*: Wings hyaline, 5th abdominal sternum mostly membranous medially, hind tibial spines in 2 or 3 rows.

FEMALE DIAGNOSIS. Medium large, 6-7 mm long; head and thoracic pleura rugulose with dense, coarse pits; clypeus black, margin sharp and distinctly upturned (Fig. 78:D); propodeal carinae weak, apophysis weak; 1st abdominal tergum moderately slender (Fig. 84:C), postpetiole moderately widened; 3rd valvula moderately long, as long as basal 4 hind tarsomeres; flagellum bicolored, moderately slender, narrowed apically (Fig. 80:E), basal 3 segments longer than wide; legs I and II yellow, leg III mostly orange.

FEMALE DESCRIPTION. *Yellow*: Scape, flagellum on basal half, tegula, legs I and II, coxae III ventrally. *Orange*: Mandible, leg III except coxa ventrally and trochanters, abdomen basally. *Black*: Flagellum on apical half, clypeus, and abdomen basally. *Punctation* (Fig. 86:F): Head rugulose with very dense, coarse pits; propleurum and mesopleurum rugulose with variably dense, coarse pits. *Shape*: Clypeal margin sharp and distinctly upturned (Fig. 78:D); flagellum moderately slender, narrowed apically, basal 3 segments longer than wide (Fig. 80:E,H); areola broadly hexagonal, area dentipara narrowed, apophysis weak but distinct, propodeal carinae weak; 1st tergum moderately slender (Fig. 84:C), postpetiole moderately and gradually widened, about 1.67 times as wide as long; femur III swollen (Fig. 79:E), ratio 0.28. *Other*: 3rd valvula moderately long, as long as basal 4 hind tarsomeres, wings slightly tinted, and hind tibial spines strong.

REMARKS. See discussion under *aurantifex*.

RANGE. Central and eastern United States and southern Canada, from Minnesota and Ontario southward to Colorado and Texas and eastward to the Atlantic Coast. Collections June and July.

MATERIAL EXAMINED. **Holotype:** male, Moorestown, NEW JERSEY, 16-vi-1939, H. and M. Townes (AEI). **Paratypes:** 21 M, 9 F (AEI, CNC, Dasch, OSU, UM, USNM, Yu): COLORADO, MICHIGAN, MINNESOTA, NEW JERSEY, OHIO, ONTARIO, SOUTH CAROLINA, TEXAS.

ETYMOLOGY. *Rugi-* (rough, wrinkled) + *ceps* (< caput, head), referring to the rugulose head.

Endasys spicus Luhman, sp. n.
(Fig. 61:A-D; Map 8)

MALE DIAGNOSIS. Medium small, 6.5 mm long; punctation of head rugulose with dense, coarse pits, propleurum mostly impunctate, smooth and shiny; 1st abdominal tergum slender (Fig. 84:G), postpetiole narrow; areola nearly hexagonal, area dentipara regular, apophysis sharp and toothlike, propodeum pale orangish; leg III and abdomen mostly pale orange, scape yellow, clypeus mostly black with distinctly upturned margin (Fig. 78:D); flagellum slender (Fig. 80:A), segments distinctly longer than wide to apex; radial cell short, about twice as long as wide, Rs straight; mostly southeastern United States.

MALE DESCRIPTION. *Yellow*: Scape, tegula, trochanters, sometimes coxa I more yellow, femora I and II, tibiae I and II. *Orange*: Clypeus apically dark orange, mandible, coxae pale orange, femur III, tibia III except apically, abdomen. *Black*: Most of clypeus, tibia III apically, hind tarsomeres; sometimes blackish on abdominal terga 3-apex. *Punctation* (Fig. 86:A,D-F): Face and frons a little rugulose with dense, coarse pits; temple a little rugulose with dense, more or less distinct pits; propleurum sparsely punctate to impunctate, more or less distinct pits; mesopleurum impunctate centrally with sparse, distinct pits; 1st and 2nd terga smooth and shiny. *Shape*: Clypeal margin distinctly upturned (Fig. 78:D); flagellum slender (Fig. 80:A), segments distinctly longer than wide to apex; areola nearly hexagonal, area dentipara regular, apophysis sharp and toothlike, propodeal carinae distinct; 1st tergum slender (Fig. 84:G), postpetiole longer than wide, 2nd tergum a little narrowed; femur III moderately swollen (Fig. 79:C), ratio 0.22-0.23; Rs straight, radial cell short, about twice as long as wide. *Other*: Wings tinted, nervulus and brace veins intercepting, 5th abdominal sternum membranous medially, and distinct cranial suture below middle ocellus.

FEMALE DIAGNOSIS. Medium small, 5-6.5 mm long; clypeal margin sharp and distinctly upturned (Fig. 78:D); 3rd valvula long, as long as hind tarsomeres; white on flagellar segments 4 or 5 to 9, flagellum linear to apex, basal 3 segments elongate (Fig. 80:F,I); area dentipara regular, apophysis sharp and toothlike, distinctly projecting; yellow scape, tegula, legs I and II, coxa III and hind trochanters; pale orange femur III, tibia III, and abdomen; petiole long and slender, postpetiole widened abruptly, sides parallel; mostly southeastern United States.

FEMALE DESCRIPTION. *White*: Flagellar segments 4 or 5 to 9. *Yellow*: Scape and pedicel, tegula, legs I and II, coxa III, hind trochanters. *Pale orange*: Mandible, femur III, tibia III, abdomen, apophysis. *Black*: Flagellum orangish black basally and apically, basal segments lighter ventrally; clypeus apically dark orange; hind tarsomeres blackish orange. *Punctation* (Fig. 86:E-F): Head rugulose with very dense, coarse pits; propleurum a little rugulose with dense, coarse pits; mesopleurum with mostly sparse, coarse pits, surface more or less rugulose. *Shape*: Clypeal margin sharp and distinctly upturned (Fig. 78:D); flagellum moderately stout, linear to apex, basal 3 segments distinctly elongate (Fig. 80:F,I); areola nearly hexagonal, distinctly elongate anteriorly, area dentipara regular, apophysis sharp and toothlike, propodeal carinae moderately distinct; 1st tergum slender (Fig. 84:C), postpetiole moderately but abruptly widened, sides more or less parallel; femur III moderately swollen (Fig. 79:D), ratio 0.25; radial cell short, about as long as 2nd-discoidal cell, Rs straight. *Other*: 3rd valvula long, as long as hind tarsomeres; wings tinted.

REMARKS. This species exhibits characters of the sister genus *Amphibulus* by the slender flagellum and 1st abdominal tergum, as well as by the toothlike apophysis and the interception of the nervulus by the brace vein. It is placed in *Endasys* because of the form of the prepectal carina, the length of the hind tibial spurs, the ridge across the prescutellar groove, and the emarginate shape of the male subgenital plate. See additional discussion under *auriculiferus*.

RANGE. Florida, Alabama, North Carolina, and New York. Collections April through July.

MATERIAL EXAMINED. **Holotype**: male, Gulf Shores, ALABAMA, 25-IV-1968, H. and M. Townes (AEI). **Paratypes**: 6 M, 2 F (AEI, USNM): ALABAMA, FLORIDA, NEW YORK, NORTH CAROLINA. Other material studied: 1 M, 2 F (FLDA): FLORIDA.

ETYMOLOGY. *Spicus* (spike), referring to the spike-like apophyses of both male and female.

SANTACRUZENSIS GROUP

DIAGNOSIS. Clypeus often impressed apically, margin distinctly upturned, propodeal carinae distinct but fine, apophysis weak or absent, 1st abdominal tergum of male slender, carinae reduced or nearly absent; Southwest United States, southern California, and northern Mexico.

Twenty species are included here, although *chiricahuanus* with reservations. These species exhibit many features of *Amphibulus*, particularly in the form of the clypeus, propodeum, and 1st abdominal tergum. However, they are generally distinguished from that genus by distinct dimorphism, the smooth sternaulus, the shape of the area dentipara, always distinct ridge across the prescutellar groove, transverse break on mesoscutum before this groove not as distinct, and flagellum of female never with annulus.

The most distinctive features of this group are the following: clypeus impressed apically, clypeal margin sharp and distinctly upturned (except in male *chiricahuanus*); face a little wider than height of face plus clypeus; male flagellum often with more than 2 tyloids, additional tyloids on 12th or 13th segments, flagellum moderately slender; female flagellum unicolored or bicolored, never with annulus; propodeum of male moderately to strongly sloped, carinae weak but distinct, areola hexagonal, male areola elongate posteriorly, female areola anteriorly, area dentipara regular to widened, apophysis reduced or absent; 1st abdominal tergum slender in male, 5th and 6th sterna membranous medially; hind femur swollen or moderately swollen; dimorphism more pronounced.

This group ranges from southern California, Arizona, and New Mexico southward to central Mexico. The only host record is for *pinidiprionis*, reared from diprionid larvae. Species are always collected in pine forests or in areas where pines grow, such as Santa Cruz Island, California.

This group corresponds to the Palearctic Senilis Group. The following Nearctic spieces are included: *arizonae, aureolus, auriger, callidius, chiricahuanus, concavus, durangensis, flavissimus, flavivittatus, gracilis, julianus, leucocnemis, melanogaster, occipitis, pinidiprionis, punctatior, santacruzensis, spinissimus, tetratylus,* and *tricoloratus.*

Endasys arizonae Luhman, sp. n.
(Fig. 31:A-D; Map 2)

MALE DIAGNOSIS. Medium size, 6-6.5 mm long; clypeus yellow, margin sharply upturned (Fig.78:E); white coxae I and II apically, tegula mostly, trochanters, and tibiae I and II dorsally; leg III black except trochanters, abdomen orange except basally, apically, and most of 2nd tergum; Arizona.

MALE DESCRIPTION. *White*: Scape except often yellowish, coxae I and II apically, trochanters ventrally, tegula. *Yellow*: Often clypeus, mandible, tibiae I and II except dorsally. *Orange*: Abdomen except basally, apically, and variably on 2nd tergum. *Black*: Flagellum, coxae I and II basally, leg III except trochanters ventrally, and abdomen basally, parameres, and often 2nd tergum. *Punctation* (Fig. 86:B,C): Face shiny, slightly rugulose or granular with dense pits; frons dense with fine pits; temple with moderately dense, indistinct pits; propleurum with dense to very sparse, very fine pits; mesopleurum impunctate centrally, otherwise with sparse, fine pits; 1st and 2nd terga smooth and shiny. *Shape*: Flagellum moderately slender (Fig. 80:B), segments longer than wide to apex; clypeal margin sharply upturned (Fig. 78:E); areola broadly hexagonal (Fig. 85:C), area dentipara narrowed, apophysis weak, propodeal carinae weak; 1st tergum moderately slender (Fig. 84:F), postpetiole longer than wide, sides slightly diverging, 2nd tergum regular; femur III moderately swollen (Fig. 79:C), ratio about 0.22. *Other*: Wings mostly clear, 5th abdominal sternum membranous medially, glumes dense (Fig. 88:A), propodeum short and nearly level dorsally.

FEMALE DIAGNOSIS. Medium size, 5.5-6.5 mm long; clypeus black with orangish black apex, margin sharply upturned (Fig. 78:E); flagellum orange, slender, narrowed apically (Fig. 80:E), basal 3 segments moniliform; legs I and II orange, coxa II blackish orange, coxa III and femur III blackish orange, hind trochanters and tibia III orange, abdomen orange except 1st tergum blackish laterally; 1st tergum moderately stout (Fig. 84:B), postpetiole moderately widened; area dentipara narrowed, apophysis weak; femur III swollen (Fig. 79:E), ratio about 0.28; 3rd valvula moderately long, as long as basal 4 hind tarsomeres; wings tinted.

FEMALE DESCRIPTION. *Orange*: Flagellum, scape, mandible, tegula, legs I and II except coxa II more blackish orange, hind trochanters, tibia III, hind tarsomeres, abdomen except 1st tergum laterally. *Black*: Clypeus except more orangish black apically, coxae II and III blackish, femur III blackish, and 1st tergum laterally to postpetiole. *Punctation* (Fig. 86:B,D-F): Face slightly rugulose with moderately dense, coarse pits; frons densely, coarsely pitted; temple with variably sparse, more or less distinct pits; propleurum densely to sparsely pitted; mesopleurum sparsely, finely pitted. *Shape*: Flagellum slender, narrowed apically, basal 3 segments moniliform (Fig. 80:E,G); clypeal margin sharply upturned (Fig. 78:E); areola elongate hexagonal, area dentipara narrowed, apophysis weak; 1st tergum moderately stout (Fig. 84:B), postpetiole moderately widened, about 1.5 times as wide as long; femur III swollen (Fig. 79:E), ratio about 0.28; Rs straight, radial cell a little longer than 2nd-discoidal cell. *Other*: 3rd valvula moderately long, as long as basal 4 hind tarsomeres; wings tinted; propodeum nearly level dorsally.

REMARKS. This species is closely related to *concavus* and *melanogaster*. Male *arizonae* differs from both by its yellow clypeus. It differs further from *concavus* by the black hind leg, and from *melanogaster* by the abdominal terga 3-6 orange, never black. Female *arizonae* is distinguished from *concavus* by the orange flagellum, narrowed on apical half, basal 3 segments moniliform, and the blackish hind coxa. Females of *melanogaster* are unknown.

RANGE. Mountainous Arizona. Collections in April.

MATERIAL EXAMINED. **Holotype:** male, Cave Ck. Cyn., Chiricahua Mts., ARIZONA, 27-IV-1981, B. and C. Dasch (AEI). **Paratypes:** 89 M, 4 F (AEI, Dasch): ARIZONA. Other material studied: 63 M (Dasch): ARIZONA.

ETYMOLOGY. Arizonae (of Arizona), referring to locality of type series.

Endasys aureolus Luhman, sp. n.
(Fig. 14:A-D; Map 12)

MALE DIAGNOSIS. Medium small size, 5.5-6.5 mm long; orange legs and entire abdomen, 1st abdominal tergum slender (Fig. 84:G), postpetiole distinctly longer than wide, 2nd tergum narrowed and slightly elongate, clypeus black with sharp and distinctly upturned margin (Fig. 78:D), face rugulose, frons densely but finely punctate, propleurum very sparsely punctate, mesopleurum mostly impunctate centrally and smooth and shiny, areola nearly hexagonal, slightly elongate apically (fig. 85:B), area dentipara regular, and apophysis weak; Arizona.

MALE DESCRIPTION. *Orange*: Scape blackish orange, mandible, tegula, legs, and abdomen. *Black*: Flagellum and clypeus. *Punctation* (Fig. 86:A,B,D,F): Face rugulose with dense pits; frons with dense, fine, more or less distinct pits; temple with sparse, more or less distinct pits; propleurum with very sparse, fine pits; mesopleurum mostly impunctate, surface smooth and shiny, otherwise with sparse, finely distinct pits; postpetiole and 2nd tergum smooth and shiny. *Shape*: Clypeal margin distinctly upturned (Fig. 78:D); flagellum moderately slender (Fig. 80:B), segments longer than wide to antennal apex; areola nearly hexagonal, elongate apically (Fig. 85:B), area dentipara regular, apophysis weak, propodeal carinae distinct; 1st tergum slender (Fig. 84:G), postpetiole distinctly elongate, 2nd tergum narrowed and appearing elongate; femur III moderately swollen (Fig. 79:C), ratio about 0.23; radial cell slightly elongate, Rs straight (Fig. 82:B). *Other*: 2 prominent tyloids on either flagellar segments 11-12 (type) or 10-11 (paratype), and type with 2 additional very small tyloids on segments 10 and 13; wings clear with large, pentagonal areolet.

FEMALE DIAGNOSIS. Medium small, 5.5 mm long; clypeus black, margin sharp and distinctly upturned (Fig. 78:D); punctation sparse and coarse on frons, temple, propleurum, and mesopleurum, and very dense, coarse, and rugulose on frons; 1st abdominal tergum moderately slender (Fig. 84:F), postpetiole less than 1.5 times as wide as long; 3rd valvula moderately long, about as long as basal 4 hind tarsomeres; legs, abdomen, scape orange, and flagellum basally orange; Arizona.

FEMALE DESCRIPTION. *Orange*: Flagellum basal half, scape, mandible, tegula, legs, abdomen except basally. *Black*: Flagellum apical half, clypeus, abdomen basally. *Punctation* (Fig. 86: E-F): Face rugulose with very dense pits; frons increasingly rugulose toward scape, pits dense and coarse; temple with variably sparse, coarse pits; propleurum rugulose with variably sparse, coarse pits; mesopleurum very slightly rugulose with variably sparse pits. *Shape*: Clypeus with distinctly upturned margin (Fig. 78:D); flagellum moderately slender, narrowed apically, basal 3 flagellomeres short (Fig. 80:E,H); areola elongate hexagonal (Fig. 85:A), area dentipara a little narrowed and elongate apically, apophysis moderately distinct, propodeal carinae distinct; 1st tergum moderately slender (Fig. 84:C), postpetiole moderately and gradually widened, less than 1.5 times as wide as long; femur III moderately swollen (Fig. 79:D), ratio 0.25-0.26. *Other*: 3rd valvula moderately long, about as long as basal 4 hind tarsomeres.

REMARKS. See discussion under *tetratylus* and *julianus*.

RANGE. Arizona. Collections August and September.

MATERIAL EXAMINED. **Holotype**: male, Portal, ARIZONA, 7-ix-1974, H. and M. Townes (AEI). **Paratypes**: 1 M, 1 F (AEI): ARIZONA.

ETYMOLOGY. *Aureo-* (gold) + *-lus* (diminutive suffix), referring to the size and color.

Endasys auriger Luhman sp. n.
(Fig. 17:A-D; Map 1)

MALE DIAGNOSIS. Medium large, 6-8 mm long; tegula and trochanters (apically) white, yellow to yellowish orange legs I and II, tibiae I and II whitish dorsally, abdomen and most of leg III orange, hind tarsomeres black; face finely granular or finely rugulose with very dense, fine pits; clypeus impressed apically with more or less distinctly upturned margin (Fig. 78:D); temple densely punctate with more or less distinct pits; propodeal carinae weak, apophysis weak, but distinct; wings clear; 1st abdominal tergum moderately slender (Fig. 84:F), postpetiole squarish; femur III moderately swollen (Fig. 79:C), ratio about 0.22; Arizona.

MALE DESCRIPTION. *White*: tegula, trochanters apically, tibiae I and II dorsally. *Yellow*: Tyloids at base, scape, legs I and II except tibiae dorsally and trochanters apically. *Orange*: Usually clypeus, mandible, abdomen except often basally, femur III, tibia III except dorso-apically. *Black*: Most of flagellum, sometimes clypeus more blackish, usually tibia III dorso-apically, hind tarsomeres, often abdomen basally; dark forms coxae orangish black, femur III with blackish, and tibia III basally and apically. *Punctation* (Fig. 86:A-B,D-F): Face finely granular or finely rugulose with very dense, fine pits; frons very densely pitted; temple with moderately dense, more or less distinct pits, surface a little rugulose; propleurum with dense to sparse, distinct pits and some rugulosity; mesopleurum impunctate centrally, otherwise variably dense to sparse with distinct pits and some rugulosity; 1st and 2nd terga smooth and shiny. *Shape*: Clypeus impressed apically with more or less distinctly upturned margin (Fig. 78:C); flagellum moderately stout (Fig. 80:C), segments squarish from tyloids to apex; areola broadly to elongate hexagonal, area dentipara narrowed, apophysis weak but distinct, propodeal carinae weak, propodeum a little sloped apically; 1st tergum moderately slender (Fig. 84:F), postpetiole squarish, 2nd tergum a little widened; femur III moderately swollen (Fig. 79:C), ratio about 0.22. *Other*: Wings clear, 5th and 6th abdominal sterna membranous medially, hind tibial spines in about 3 rows.

FEMALE DIAGNOSIS. Medium, 6.5-8 mm long; mostly orange legs and abdomen, tibiae I and II pale yellow dorsally; head rugulose with very dense, coarse pits; propodeum with distinct carinae, area dentipara narrowed, apophysis moderately distinct; flagellum slender but more or less linear to apex (Fig. 80:F); 1st abdominal tergum moderately slender (Fig. 84:C), postpetiole moderately expanded, about 1.5 times as wide as long; 3rd valvula moderately short, about as long as basal 3 hind tarsomeres; femur III moderately swollen (Fig. 79:C), ratio about 0.26; Rs straight, wings clear.

FEMALE DESCRIPTION. *Yellow-orange*: Tegula, legs I and II except tibiae pale yellow dorsally, hind trochanters. *Orange*: Flagellum but paler on basal half, scape, clypeus more blackish orange, leg III except trochanters, abdomen. *Black*: Sometimes clypeus more orangish black. *Punctation* (Fig. 86: E-F): Face and frons rugulose with very dense, coarse pits; temple more or less rugulose with moderately dense, coarse pits; propleurum rugulose with dense pits; mesopleurum rugulose with variably sparse, coarse pits. *Shape*: Clypeus impressed apically with more or less distinctly upturned margin (Fig. 78:C); flagellum slender but more or less linear to apex, basal 3 segments longer than wide (Fig. 80:F,H); areola broadly hexagonal, area dentipara narrowed and high, apophysis distinct, propodeal carinae distinct; 1st tergum moderately slender (Fig. 84:C), postpetiole moderately widened, about 1.5 times as wide as long; femur III moderately swollen (Fig. 79:D), ratio 0.26-0.27; Rs straight. *Other*: 3rd valvula moderately short, about as long as basal 3 hind tarsomeres; wings more or less clear.

REMARKS. See discussion under *tetratylus*.

RANGE. Arizona. Collections July and August.

MATERIAL EXAMINED. **Holotype**: male, Parker Cyn. L., ARIZONA, 22-viii-1974, H. and M. Townes (AEI). **Paratypes**: 26 M, 3 F (AEI. USNM): ARIZONA. Other material studied: 1 M (AEI): ARIZONA.

ETYMOLOGY. *Auri-* (gold) + *-ger* (bearer), referring to the golden orange of the legs and abdomen.

Endasys callidius Luhman, sp. n.
(Fig. 10:A-D; Map 7)

MALE DIAGNOSIS. Small, about 3-5 mm long; abdomen shiny black, coxae white at least apically, white trochanters, scape, and clypeus, legs mostly black except white trochanters and yellow tibiae basally; frons densely to sparsely punctate with fine, more or less distinct pits; 1st abdominal tergum moderately slender (Fig. 84:F), postpetiole longer than wide, 2nd tergum regular; femur III moderately swollen (Fig. 79:C), ratio 0.22-0.25;

area dentipara narrowed, apophysis weak; southeastern Arizona and San Diego Co., California.

MALE DESCRIPTION. *White*: Scape, clypeus, mandible, tegula, coxae I and II except usually coxa II basally, coxa III apically, trochanters. *Yellow*: Flagellum ventrally, femora I and II, tibiae I and II except pale yellow dorsally, usually coxa II basally, coxa III baso-ventrally, sometimes femur III basally and ventrally, tibia III basally. *Orange*: Sometimes abdomen more orange on 2nd tergum apically and terga 3-6 basally. *Black*: Flagellum dorsally, coxa III dorso-basally, femur III except sometimes basally and ventrally, tibia III apically, hind tarsomeres, abdomen except sometimes more orange on 2nd tergum apically and terga 3-6 basally. *Punctation* (Fig. 86:A-E): Face finely rugulose with dense pits; frons with variably dense to sparse punctation with fine, more or less distinct pits; temple with moderately sparse, indistinct pits; propleurum with mostly sparse, very fine, more or less distinct pits; mesopleurum impunctate centrally, otherwise with sparse, more or less distinct pits; postpetiole and 2nd tergum mostly smooth and shiny. *Shape*: Clypeal margin not upturned (Fig. 78:B); flagellum moderately slender, segments longer than wide to apex (Fig. 80:B); areola broadly hexagonal (Fig. 85:C), area dentipara narrowed, apophysis moderately weak, propodeum shiny between moderately strong carinae; 1st tergum moderately slender (Fig. 84:F), postpetiole longer than wide, 2nd tergum regular or squarish; femur III moderately swollen (Fig. 79:C), ratio 0.22-0.25. *Other*: wings clear, and 5th abdominal sternum membranous medially.

FEMALE DIAGNOSIS. Small, 3.5-4 mm long; face finely rugulose with dense pits; yellow clypeus mostly flat with sharp, slightly upturned margin (Fig. 78:B); frons and temple mostly sparse with more or less distinct pits; flagellum more yellowish, moderately stout, linear to apex (Fig. 80:F); yellow coxae I and II, coxa III apically, trochanters, and tibia III; mostly black coxa III and femur III; abdomen black except apically on 1st tergum and basally on terga 3-6, remaining areas orangish; femur III swollen (Fig. 79:E), ratio about 0.3; 3rd valvula moderately long, about as long as basal 4 hind tarsomeres; radial cell short, Rs curved (Fig. 82:D).

FEMALE DESCRIPTION. *Yellow*: Flagellum (darker dorso-apically), scape, clypeus, mandible, tegula, coxae I and II except basally, coxa III apically, trochanters, tibiae I and II except posteriorly, tibia III. *Orange*: 1st abdominal tergum apically and terga 2-6 except apically and laterally. *Black*: Coxa II basally, tibia II posteriorly, coxa III except apically, femur III, abdomen basally and apically, basally and laterally on 1st tergum, apically and laterally on terga 3-6 and laterally on terga 3-6. *Punctation* (Fig. 86:B,D,F): Face finely rugulose with dense pits; frons with mostly sparse, more or less distinct pits; temple with sparse, fine pits; propleurum with mostly sparse, fine pits; mesopleurum with mostly sparse punctation and slight rugulosity with more or less distinct pits. *Shape*: Clypeus mostly flat, margin sharp but only slightly upturned; flagellum moderately stout, linear to apex, basal 3 segments short (Fig. 80:F,H); areola broadly hexagonal, area dentipara narrowed, apophysis moderately weak; 1st tergum moderately slender (Fig. 84:C), postpetiole squarish; femur III swollen (Fig. 79:E), ratio about 0.3; radial cell short, Rs curved (Fig. 82:D). *Other*: 3rd valvula moderately long, about as long as basal 4 hind tarsomeres; wings more or less clear.

REMARKS. This species is not clearly related to other species in the Santacruzensis Group. It lacks a distinctly upturned clypeal margin, but is placed in this group because of the color pattern, propodeal characters, and the 1st abdominal tergum. It may be related to *tricoloratus,* from which males are distinguished by the black femur III and mostly black abdomen. Females differ by the blackish coxa III, black femur III, and yellow flagellum.

RANGE. Southeastern Arizona and southern San Diego Co., California. Collections April through June and August through September.

MATERIAL EXAMINED. **Holotype**: male, Oak Ck. Cyn., ARIZONA, 21-v-1947, H. and M. Townes (AEI). **Paratypes**: 13 M, 2 F (AEI, UCR): ARIZONA, CALIFORNIA

ETYMOLOGY. (Greek) *Call-* (beauty) + *-idius* (diminutive suffix), referring to the small size and pretty combination of colors.

Endasys chiricahuanus Luhman, sp. n.
(Fig. 58:A-D; Map 7)

MALE DIAGNOSIS. Medium small, 5-6 mm long; clypeus black, margin not upturned (Fig. 78:B); face densely punctate, appearing finely granular, frons densely and finely punctate; coxae mostly black, trochanters yellow, femora I and II and tibiae I and II yellow, femur III and tibia III orange with black apices, abdomen orange with black basally and apically, and scape yellowish; flagellum moderately stout (Fig. 80:C), segments more or less square on apical half; propodeal carinae moderately strong, apophysis distinct; 1st tergum moderately slender (Fig. 84:F), postpetiole longer than wide; femur III moderately swollen (Fig. 79:C), ratio about 0.23; wings very slightly tinted; Arizona.

MALE DESCRIPTION. *Yellow*: Scape, coxae I and II apically, trochanters, femora I and II, tibiae I and II except latter pale yellow dorsally. *Orange*: Mandible apically, femur III and tibia III except apices, abdomen except basally and apically. *Black:* Mandible basally, flagellum except often lighter ventrally, clypeus except often orangish black, tegula except sometimes more orangish, coxae I and II except apically, coxa III, femur III, tibia III apically, hind tarsomeres, abdomen basally and apically; sometimes blackish apically on terga 4-6. *Punctation* (Fig. 86:A-F): Face finely granular or finely rugulose with dense pits; frons with dense, finely distinct pits; temple with dense, indistinct pits; propleurum with dense, indistinct pits, to impunctate; mesopleurum impunctate centrally, otherwise with sparse, distinct pits; 1st and 2nd terga smooth and shiny. *Shape*: Clypeal margin not upturned (Fig. 78:B); flagellum moderately stout (Fig. 80:C), segments squarish on apical half; areola broadly hexagonal, area dentipara narrowed, apophysis distinct, propodeal carinae moderately strong; 1st tergum moderately slender (Fig. 84:F), postpetiole longer than wide, 2nd tergum regular; femur III moderately swollen (Fig. 79:C), ratio about 0.23; Rs more or less straight. *Other*: Wings slightly tinted, 5th abdominal sternum membranous medially, hind tibial spines in about 3 rows.

FEMALE DIAGNOSIS. Small, about 4 mm long; clypeus blackish, margin very slightly upturned (Fig. 78:B); face more or less rugulose with dense pits, frons densely punctate with fine pits, temple with moderately dense, more or less distinct pits; flagellum more or less bicolored, moderately stout, linear to apex (Fig. 80:F); yellow scape, tegula, and legs except coxa III more blackish, abdomen orange; 1st tergum moderately stout (Fig. 84:B), postpetiole moderately widened, about 1.5 times as wide as long; femur III moderately swollen (Fig. 79:C), ratio about 0.27; 3rd valvula moderately long, as long as basal hind tarsomeres; Rs straight.

FEMALE DESCRIPTION (based on one specimen). *Yellow*: Flagellum basal half, scape, mandible, tegula, and legs except coxa III. *Orange*: Abdomen except 3rd valvula. *Black*: Flagellum more yellowish black on apical half, clypeus yellowish black, coxa III yellowish black, 3rd valvula. *Punctation* (Fig. 86:B,D,F): Face more or less finely rugulose with dense pits; frons with dense, fine pits; temple with moderately dense, more or less distinct pits; propleurum with moderately dense, very fine pits; mesopleurum finely rugulose, with variable punctation. *Shape*: Clypeal margin very slightly upturned (Fig. 78:B); flagellum moderately stout, linear to apex, basal 3 segments short (Fig. 80:F,H); areola broadly hexagonal, area dentipara narrowed and a little elongate apically, apophysis distinct and projecting a little obliquely, propodeal carinae distinct; 1st tergum moderately stout (Fig. 84:B), postpetiole moderately widened, about 1.5 times as wide as long; femur III moderately swollen (Fig. 79:D), ratio about 0.27; Rs straight. *Other*: 3rd valvula moderately long, as long as basal 4 hind tarsomeres; wings lightly tinted.

REMARKS. This species is included in the Santacruzensis Group with reservations because the clypeal margin is scarcely upturned. It is placed here primarily because of the color pattern, the shape of the 1st abdominal tergum, and the distribution. Other species in this group with a similar color pattern differ by having a distinctly upturned clypeal margin, or the coxae have white at least apically.

RANGE. Mountainous Arizona and New Mexico. Collections April and May, and July through October.

MATERIAL EXAMINED. **Holotype**: male, Cave Ck. Cyn., Chiricahua Mts. ARIZONA, 27-iv-1981, B. and C. Dasch (AEI). **Paratypes**: 12 M, 1 F (AEI, CAS, Dasch): ARIZONA, NEW MEXICO.

ETYMOLOGY. From the Chiricahua Mts., the type locality.

Endasys concavus Luhman, sp. n.
(Fig. 30:A-D; Map 6)

MALE DIAGNOSIS. Medium large size, 6-8 mm long; clypeus black, margin sharply upturned (Fig. 78:E); propodeal carinae weak, apophysis absent or weak; 1st abdominal tergum slender (Fig. 84:G), postpetiole distinctly elongate, 2nd tergum narrowed; coxae mostly blackish white on coxae I and II apically, trochanters white, femur III orange, tibia III orange with black apex, abdomen orange except basally and apically, yellowish scape and tegula; Arizona.

MALE DESCRIPTION. *White*: Coxae I and II apically, trochanters except hind 1st-trochanter. *Yellow*: Scape at least apically, tegula, femora I and II, tibiae I and II, latter pale yellow dorsally. *Orange*: Mandible sometimes more orangish black, femur III, tibia

III except apically and sometimes basally, coxa III ventrally, abdomen except basally and apically. *Black*: Scape at least basally, clypeus, usually most of mandible, usually coxae I and II except apically, coxa III ventrally, hind 1st-trochanter, tibia III apically and often basally, hind tarsomeres, abdomen basally and apically, sometimes tergum 2 black laterally and apically. *Punctation* (Fig. 86:A-B,D-E): Face finely granular with dense pits; frons densely pitted; temple with moderately dense, more or less distinct pits; propleurum with dense to very sparse, finely distinct pits; mesopleurum impunctate centrally, otherwise sparsely pitted; 1st and 2nd abdominal terga smooth and shiny. *Shape*: Clypeus impressed apically, margin sharp and distinctly upturned (Fig. 78:E); flagellum moderately slender (Fig. 80:B), segments longer than wide to apex; areola elongate hexagonal (Fig. 85:B), area dentipara regular, apophysis absent or weak, propodeal carinae weak; 1st tergum slender (Fig. 84:G), postpetiole distinctly elongate, 2nd tergum narrowed; femur III moderately slender (Fig. 79:B), ratio 0.21. *Other*: Wings more or less clear or slightly tinted; 5th abdominal sternum membranous medially; glumes moderately dense (Fig. 88:B).

FEMALE DIAGNOSIS. Medium large, 6 mm long; flagellum bicolored, linear apical half (Fig. 80:F), basal 3 segments a little elongate; coxae black, tibiae I and II pale yellow dorsally; femur III moderately swollen (Fig. 79:C), ratio 0.25; 3rd valvula moderately short, about as long as basal 3 hind tarsomeres; 1st tergum moderately stout (Fig. 84:B), postpetiole moderately wide, sides nearly parallel; clypeus black, margin sharply upturned (Fig. 78:E); face and frons densely and distinctly punctate, surfaces with rugulosity; areola broadly hexagonal, area dentipara narrowed, apophysis moderately distinct, propodeal carinae distinct; radial cell mostly straight.

FEMALE DESCRIPTION. *Pale yellow*: Tibiae I and II dorsally. *Orange*: Scape, flagellum on basal half, legs except coxae and tibiae I and II dorsally, abdomen except basally, 3rd valvula. *Black*: Flagellum apical half, tegula, coxae, abdomen basally, 3rd valvula. *Punctation* (Fig. 86:D-E): Face and frons finely rugulose with dense pits; temple densely pitted; propleurum slightly rugulose with dense pits; mesopleurum sometimes with punctation centrally, otherwise with variably dense, variably distinct pits, surface slightly rugulose. *Shape*: Clypeus impressed apically, margin distinctly upturned (Fig. 78:E); flagellum moderately stout, segment linear in apical half, basal 3 segments elongate (Fig. 80:F,I); areola broadly hexagonal, area dentipara narrowed, apophysis moderately distinct, propodeal carinae distinct; 1st tergum moderately stout (Fig. 84:B), postpetiole gradually widened to about 1.67 times as wide as long; femur III moderately swollen (Fig. 79:D), ratio 0.25; Rs mostly straight. *Other*: 3rd valvula moderately short, as long as basal 3 hind tarsomeres; wings very slightly tinted.

REMARKS. The female description is based on one specimen. Association with the male is made primarily by the following correspondences: color patterns of the coxae, tibiae dorsally, and femur III; punctation of the head; and similarity of locality—Arizona mountains. Furthermore, the only other similar species in the locality has a clearly associated female. See additional discussion under *arizonae*.

RANGE. Arizona. Collections April, May, and early July.

MATERIAL EXAMINED. **Holotype**: male, Parker Ck., Sierra Ancha, ARIZONA, 9-v-1947, H. and M. Townes (AEI). **Paratypes**: 28 M, 1 F (AEI, Dasch, UCR): ARIZONA. ETYMOLOGY. *Concavus* (concave), referring to the impressed clypeus.

Endasys durangensis Luhman, sp. n.
(Fig. 75:A-D; Map 7)

MALE DIAGNOSIS. Medium large, 6.5-7.5 mm long; face rugulose with very dense, coarse pits; clypeus black, densely and coarsely punctate, margin more or less upturned (Fig. 78:C); propodeum weakly carinate, more or less sloped, area dentipara nearly regular, apophysis weak and inconspicuous; frons irregularly and densely punctate, temple with punctation moderately dense, pits more or less distinct; tegula white, coxae mostly black, trochanters yellow, femur III mostly orange except apically black, tibia III orange with black dorso-apically, abdomen shiny and orange with black basally and parameres; 5th abdominal sternum membranous medially; 1st tergum smooth and shiny, longer than wide; tibiae I and II pale yellow dorsally; central, montane Mexico.

MALE DESCRIPTION. *White*: tegula. *Yellow*: Scape, clypeus sometimes (paratypes), legs I and II except most of coxae I and II pale dorsally, trochanter III except more orangish basally. *Orange*: Mandible apically, trochanter III basally, femur III of type basally and dorsally, femur III of paratypes mostly orange except apically, tibia III except dorso-apically, abdomen except basally and apically. *Black*: Clypeus (type), mandible basally, most of coxae I and II except apically, coxa III, most of femur III of type except basally and dorsally, femur III of paratypes apically, tibiae III dorso-apically, hind tarsomeres, abdomen basally and apically, often only parameres. *Punctation* (Fig. 86:A-B,D-F): Face and clypeus rugulose with very dense, coarse pits; frons with dense, distinct pits, sometimes slightly rugulose; temple with moderately dense, more or less distinct pits; propleurum with dense to very sparse, fine pits; mesopleurum impunctate centrally, otherwise with sparse, more or less distinct pits; 1st and 2nd terga smooth and shiny. *Shape*: Clypeal margin more or less upturned (Fig. 78:C); flagellum moderately stout (Fig. 80:C), segments squarish on apical half; areola irregularly hexagonal, elongate posteriorly (Fig. 85:B), area dentipara nearly regular, apophysis weak, propodeum more or less sloped, carinae weak; 1st tergum moderately slender (Fig. 84:F), postpetiole a little longer than wide, 2nd regular; femur III moderately slender (Fig. 79:B), ratio about 0.20; radial cell straight, Rs slightly bowed (Fig. 82:A). *Other*: Wings clear; 5th abdominal sternum membranous medially; tibia III slender on basal 0.67, more abruptly widened on apical third; glumes dense (Fig. 88:A).

FEMALE DIAGNOSIS. Medium large, 6-7 mm long; face and frons rugulose with very dense and coarse pits, temple a little rugulose with densely coarse pits; yellow tegula, legs I and II with tibiae pale dorsally; flagellum bicolored, segments 4 or 5-9 paler, more or less linear to apex, basal 3 segments elongate (Fig. 80:F,I); orange leg III and abdomen; 3rd valvula black, as long as basal 2 hind tarsomeres; Rs straight; central, montane.

FEMALE DESCRIPTION. *Yellow*: Tegula, legs I and II except tibiae pale dorsally. *Orange*: Flagellum basal half, segments 4 or 5 to 9 paler, mandible, hind leg, and abdomen except 3rd valvula. *Black*: Flagellum apical half, clypeus, and 3rd valvula. *Punctation* (Fig. 86:B,E-F): Face and frons a little rugulose with very dense pits; temple a little rugulose with dense, coarse pits; propleurum with dense, fine pits, surface finely rugulose; mesopleurum rugulose, impunctate centrally, otherwise with variably dense to sparse pits. *Shape*: Clypeus more or less impressed apically, margin more or less upturned (Fig. 78:C); flagellum moderately stout, more or less linear to apex, basal 3 segments elongate (Fig. 80:F,I); areola broadly hexagonal, area dentipara distinctly narrowed, apophysis distinct, projecting at an angle posteriorly, propodeal carinae weak; 1st tergum moderately stout (Fig. 84:B), postpetiole moderately widened, less than twice as wide as long; femur III moderately swollen (Fig. 79:D), ratio 0.25-0.26; Rs straight. *Other*: 3rd valvula short, as long as basal 2 hind tarsomeres; wings very lightly tinted.

REMARKS. For affinities, see remarks under *occipitis, pinidiprionis,* and *spinissimus*.

RANGE. Central Mexico. Collections in July.

MATERIAL EXAMINED. **Holotype**: male, San Antonio, El Salto, DURANGO, 3-vii-1964, J.F. McAlpine (CNC). **Paratypes**: 2 M, 2F (CAS, CNC): DURANGO, MEXICO CITY.

ETYMOLOGY. From Durango, the Mexican state of the type locality.

Endasys flavissimus Luhman, sp. n.
(Fig. 77:A-D; Map 1)

MALE DIAGNOSIS. Medium large, 6 mm long; mostly yellow legs and abdomen except blackish tibia III basally and apically, hind tarsomeres, and abdomen basally, apically, some blackish yellow laterally and apically on terga 2-7; scape and clypeus yellow, tegula white; clypeal margin very slightly upturned (Fig. 78:C); face granular or finely rugulose with very dense pits; flagellum moderately slender (Fig. 80:B), segments elongate; propodeum smooth and shiny between moderately strong carinae, apophysis distinct; 1st tergum slender (Fig. 84:G), postpetiole elongate; wings clear; surfaces of head and thoracic pleura smooth and shiny; northern Mexico.

MALE DESCRIPTION (based on one specimen). *White*: Tegula. *Yellow*: Tyloids, scape, clypeus, mandible, legs I and II except tibiae dorsally, leg III except tibia basally and hind tarsomeres, abdomen generally except basally, parameres, and laterally and apically on terga 2-7. *Black*: Flagellum, tibia III basally and apically, hind tarsomeres, abdomen basally, parameres, and blackish yellow laterally and apically on terga 2-7. *Punctation* (Fig. 86:A-E): Face granular or finely rugulose with very dense pits; frons with dense, distinct pits; temple with moderately dense, more or less indistinct pits; propleurum sparse to impunctate, pits very fine, surface smooth and shiny; mesopleurum mostly impunctate centrally, otherwise with sparse, very fine pits, surface smooth and shiny; 1st and 2nd terga smooth and shiny. *Shape*: Clypeal margin very slightly upturned (Fig. 78:B); flagellum moderately slender (Fig. 80:B), segments distinctly longer than wide;

areola broadly hexagonal, elongate posteriorly (Fig. 85:B), area dentipara nearly regular, apophysis distinct, propodeum mostly smooth and shiny between moderately strong carinae; 1st tergum long and slender (Fig. 84:G), postpetiole elongate, 2nd tergum regular; femur III moderately slender (Fig. 79:B), ratio 0.20; radial cell long, Rs straight. *Other*: Tyloids (2) long, nearly as long as segment (Fig. 87:A); wings clear; 5th and 6th abdominal sterna membranous medially; glumes dense to very dense; nervulus distant from brace vein by about width of latter.

FEMALE. Unknown.

REMARKS. Known only from the type. The species is distinct from other species in Mexico and Arizona by the combination of color, punctation, antennal, and abdominal features. See additional remarks under *pinidiprionis*.

HOLOTYPE. Male, N. Bochil, 6500 ft., CHIAPAS, 24-vi-1969, W.R. Mason (CNC).

ETYMOLOGY. *Flav-* (yellow) + *-issimus* (-est), referring to the mostly yellow color of the legs and abdomen.

Endasys flavivittatus Luhman, sp. n.
(Fig. 73:A-D; Map 2)

MALE DIAGNOSIS. Medium small, 4.5-6.5 mm long; white coxae I and II, white trochanters, scape and clypeus black, coxa III black basally with orange ventrally, femur III orange with black dorsally, abdomen generally black with apical margins of terga 1-6 yellow; 1st tergum distinctly slender (Fig. 84:G); femur III slender (Fig. 79:A), ratio 0.21; California.

MALE DESCRIPTION. *White*: Coxae I and II except basally, trochanters I and II, hind trochanters ventrally. *Yellow*: Mandible, coxae I and II basally, coxa III laterally, apical margins of abdominal terga 1-6; sometimes most of legs I and II mostly yellow. Orange: Clypeus sometimes apically dark orange, sometimes scape more orangish, coxa III ventrally, femur III except dorsally, tibia III basally, sometimes variably orangish on terga 2-6 basally. *Black*: Usually scape, clypeus except sometimes apically, coxa III basally, femur III dorsally, tibia III except basally, hind tarsomeres, most of abdomen except apical margins of terga 1-6. *Punctation* (Fig. 86:A-C): Face finely granular; frons and temple with moderately dense, very fine, more or less distinct pits; propleurum with dense punctation along upper margin, otherwise mostly impunctate, surface smooth and shiny; mesopleurum impunctate centrally, otherwise with sparse and indistinct pits; 1st and 2nd terga smooth and shiny. *Shape*: Flagellum moderately slender (Fig. 80:B), segments longer than wide to apex; clypeal margin more or less upturned (Fig. 78:C); areola hexagonal, elongate posteriorly, area dentipara wide, apophysis weak or indistinct, propodeal carinae weak; 1st tergum distinctly slender (Fig. 84:G), postpetiole elongate, 1.5 times longer than wide, 2nd tergum narrowed; femur III slender (Fig. 79:A), ratio 0.20-0.21; Rs straight. *Other*: Tyloids (2) moderately long (Fig. 87:B), wings clear, 5th abdominal sternum complete.

FEMALE DIAGNOSIS. Small, 4.5-5.5 mm long; areola distinctly elongate, longer than wide, area dentipara narrowed, apophysis weak; 3rd valvula very short, about as long as basal 2 hind tarsomeres; radial cell short and curved; flagellum bicolored, linear to apex, basal 3 segments nearly moniliform (Fig. 80:F,G); femur III orangish or with some blackish laterally; abdomen mostly orange except basally, or sometimes blackish laterally; clypeus black basally, orange apically, margin distinctly upturned (Fig. 78:D).

FEMALE DESCRIPTION. *Yellow*: Tegula, legs I and II, trochanters III. *Orange*: Flagellum basal half, scape, clypeus apically, mandible, coxa III except basally, femur III, tibia III except apically, abdomen except basally and sometimes laterally. *Black*: Flagellum apical half, clypeus basally, coxa III basally, tibia III apically blackish, hind tarsomeres blackish, sometimes femur III blackish laterally, abdomen basally and sometimes laterally. *Punctation* (Fig. 86:C-E): Face rugulose with very dense pits; frons with dense, fine, more or less distinct pits; temple with sparse, more or less distinct pits; propleurum with mostly sparse, fine, indistinct pits; mesopleurum with very sparse, fine, indistinct pits, surface a little rugulose. *Shape*: Flagellum moderately stout, linear to apex, basal 3 segments nearly moniliform (Fig. 80:F-G); clypeal margin distinctly upturned (Fig. 78:E); areola elongate hexagonal, distinctly longer than wide, area dentipara narrowed, apophysis weak, propodeal carinae weak; 1st tergum moderately stout (Fig. 84:B), postpetiole moderately widened; femur III swollen (Fig. 79:E), ratio 0.29-0.32; radial cell short, Rs curved (Fig. 82:D). *Other*: 3rd valvula distinctly short, about as long as basal 2 hind tarsomeres; wings clear.

REMARKS. See discussion under *gracilis*.

RANGE. Coastal central and southern California. Collections April and May.

MATERIAL EXAMINED. **Holotype**: male, Potrero, San Diego Co., CALIFORNIA, 13-iv-1974, H. and M. Townes (AEI). **Paratypes**: 50 M, 44 F (AEI, CAS, OSU, UCB): CALIFORNIA. Other material studied: 9 M, 2 F (AEI, CNC, OSU): CALIFORNIA.

ETYMOLOGY. *Flavi-* (yellow) + *vittatus* (banded), referring to the yellow apical bands of the male abdominal terga.

Endasys gracilis Luhman, sp. n.
(Fig. 12:A-D; Map 5)

MALE DIAGNOSIS. Medium small, 5-6.5 mm long; 1st abdominal tergum very slender (Fig. 84:G), 2nd tergum narrowed; propodeum elongate, areola and area dentipara long, apophysis weak or lacking; white scape, clypeus, coxae I and II, coxa III apically, and trochanters; abdomen mostly blackish with orangish basally on terga 3-5, and yellowish on apical margin of terga 1-4; punctation of frons dense with fine pits; coastal California from San Francisco to San Diego Co.

MALE DESCRIPTION. *White*: Scape, clypeus except sometimes more yellow, mandible, tegula, coxae I and II, coxa III apically, trochanters. *Yellow*: Sometimes clypeus, tibiae I and II, and apical margin of terga 1-4. *Orange*: Sometimes coxa III ventrally, femur III except apically, tibia III ventrally, and abdominal terga 3 to 5 or 6. *Black*: Flagellum, coxa

III basally, femur III apically, tibia III dorsally more orangish black, hind tarsomeres, abdomen basally and apically and variably on terga 2-5 apically and laterally, sometimes abdomen entirely black except apical margins of terga 1 to 3 or 4. *Punctation* (Fig. 86:A-C): Face mostly smooth with dense, very fine pits; frons finely granular with dense, very fine pits; temple with dense, indistinct pits; propleurum with dense to very sparse punctation, or impunctate, surface smooth and shiny; mesopleurum impunctate centrally, otherwise with sparse, indistinct pits; 1st and 2nd terga smooth. *Shape*: Flagellum moderately slender (Fig. 80:B), segments longer than wide to apex; clypeus oval, margin more or less upturned (Fig. 78:C); areola nearly hexagonal, elongate apically (Fig. 85:B), area dentipara wide, propodeum slightly elongate, apophysis very weak or absent, propodeal carinae weak; 1st tergum distinctly slender (Fig. 84:B), nearly parallel-sided, postpetiole elongate, nearly twice as long as wide, 2nd tergum distinctly narrowed; femur III moderately slender (Fig. 79:B), ratio 0.19-0.21. *Other*: Tyloids (2) long, wings clear, and 5th abdominal sternum complete medially.

FEMALE DIAGNOSIS. Small, 4-5 mm long; abdominal terga 2-5 mostly orangish with at least apical and lateral areas blackish, black basally and apically; flagellum orange, moderately stout, very slightly narrowed apically, basal 3 segments nearly moniliform (Fig. 80:G); clypeus blackish orange with margin distinctly upturned (Fig. 78:E); 1st abdominal tergum moderately stout and gradually widened, sides slightly diverging; radial cell short, Rs curved; 3rd valvula distinctly short, as long as basal 2 hind tarsomeres.

FEMALE DESCRIPTION. *Yellow*: Tegula, legs I and II, hind trochanters, abdominal terga 1 and 2. *Orange*: Flagellum, scape, clypeus blackish orange, mandible, femur III basally and apically, tibia III, 1st tergum apically, variably on terga 2-4 except laterally, terga 5-7 basally, sometimes only terga 2 and 3 orangish. *Black*: Coxa III orangish black; femur III mostly orangish black except basally and apically; abdomen basally and apically, terga 2-4 at least laterally, and terga 5-apex apically and laterally, often apical half blackish. *Punctation* (Fig. 86:B-D): Face finely rugulose, frons variably dense to sparse with fine pits; temple with sparse, fine, less distinct pits; propleurum with variably sparse to dense punctation, fine, more or less distinct pits; mesopleurum with very sparse, fine, more or less distinct pits. *Shape*: Clypeal margin distinctly upturned (Fig. 78:E); flagellum moderately stout, very slightly narrowed apically, basal 3 segments nearly moniliform (Fig. 80:G); areola broadly to nearly hexagonal, area dentipara nearly regular, apophysis moderately weak, propodeal carinae very weak; 1st tergum moderately slender (Fig. 84:C), postpetiole moderately widened, about 1.5 times as wide as long; femur III swollen (Fig. 79:E), ratio 0.31-0.33. *Other*: 3rd valvula distinctly short, as long as basal 2 hind tarsomeres; wings tinted; radial cell short, Rs curved (Fig. 82:D).

REMARKS. Related to *flavivittatus*, males are distinguished by the white or yellow clypeus and the white coxae. Females are separated by the orange flagellum, black coxa III and femur III, and the longer ovipositor, about as long as the basal 3 hind tarsomeres. It is most closely related to a new Palearctic species in southern Europe. Males differ mostly by the orange femur III with black apically, females by the orange flagellum and short 3d valvula.

RANGE. Coastal California from San Diego Co. and Santa Cruz Island north to Mendocino Co. Collections March through June.

MATERIAL EXAMINED. **Holotype**: male, Potrero, San Diego Co., CALIFORNIA, 25-IV-1974, H. and M. Townes (AEI). **Paratypes**: 63 M, 70 F (AEI, UCB, UCR): CALIFORNIA. Other material studied: 5 M, 2 F (CNC, UCD, UCR): CALIFORNIA.

ETYMOLOGY. *Gracilis* (slender), referring to the slender abdomen and hind legs of the male.

Endasys julianus Luhman, sp. n.
(Fig. 16:A-D; Map 1)

MALE DIAGNOSIS. Large, 7-8 mm long; 3 short tyloids on flagellar segments 10-12; clypeus black, margin sharp and distinctly upturned (Fig. 78:D); propodeum shiny between moderately strong carinae, area dentipara narrowed, apophysis more or less distinct; 1st abdominal tergum moderately stout (Fig. 84:E), postpetiole gradually widened toward apex, apical width greater than basal width, more or less squarish; mostly orange legs and abdomen, black on femur III apically, tibia III basally and apically; wings darkened; 1st and 2nd terga smooth and shiny; southern California.

MALE DESCRIPTION. *Orange*: Scape, mandible apically, tegula, legs except femur III apically, tibia III basally and apically, abdomen except apically. *Black*: Clypeus, femur III apically, tibia III basally and apically, hind tarsus III, abdomen apically. *Punctation* (Fig. 86:A-C,E): Face finely granular with very dense, fine pits; frons with variably dense, fine pits; temple with mostly moderately sparse, indistinct pits; propleurum with dense to sparse, fine pits; mesopleurum with very sparse punctation centrally, surface smooth and shiny, otherwise with sparse, distinct pits; metapleurum shiny, pits indistinct; 1st and 2nd terga shiny and smooth. *Shape*: Clypeus with margin sharp and distinctly upturned (Fig. 78:E); flagellum moderately slender (Fig. 80:B), segments longer than wide; areola broadly hexagonal, area dentipara distinctly narrowed, apophysis more or less distinct, propodeal carinae distinct; 1st tergum moderately stout (Fig. 84:E), postpetiole gradually widened toward apex, wider apically than basally, 2nd tergum a little widened or regular; femur III moderately swollen (Fig. 79:C), ratio about 0.23. *Other*: 3 short tyloids on flagellar segments 10-12, wings darkened, 5th abdominal sternum complete apically, glumes moderately dense, and propodeum shiny between carinae.

FEMALE. Unknown.

REMARKS. This species is closely related to *aureolus*. Males differ by the very sparse punctation of central area of the mesopleurum, the square postpetiole, distinct apophysis, wings slightly infuscate, and flagellum with 3 tyloids. Females of *julianus* are unknown. See additional remarks under *tetratylus*.

RANGE. Julian, in San Diego Co., California. Collections late May.

MATERIAL EXAMINED. **Holotype**: male, Julian, San Diego Co., CALIFORNIA, 24-v-1974, H. and M. Townes (AEI). **Paratypes**: 16 M (AEI): CALIFORNIA.

ETYMOLOGY. Julian, California, referring to the type locality.

Endasys leucocnemis Luhman, sp. n.
(Fig. 56:A-D; Map 12)

MALE DIAGNOSIS. Large, 7-7.5 mm long; tibiae I and II white dorsally; clypeus black with slightly upturned margin (Fig. 78:C); face rugulose with very dense, fine pits; flagellum moderately slender (Fig. 80:B), segments slightly longer than wide on apical half, 2 short tyloids; black coxae and trochanters, femur III, and tibia III; femur III moderately swollen (Fig. 79:C), ratio 0.22; 1st tergum slender (Fig. 84:G), 2nd tergum narrowed; area dentipara nearly regular, apophysis distinct; wings a little tinted; southwestern United states.

MALE DESCRIPTION. *White*: Tibiae I and II dorsally. *Blackish yellow*: trochantelli, femora I and II, tibiae I and II except latter dorsally. *Orange*: Abdomen dark orange except basally and apically. *Black*: Flagellum, scape, mandible, tegula, coxae, 1st-trochanters, femur III, tibia III, hind tarsomeres, and abdomen basally and apically. *Punctation* (Fig. 86:A-C,F): Face rugulose with very dense pits; frons slightly rugulose with dense, fine pits; temple slightly rugulose with dense, indistinct pits; propleurum with dense to sparse, very fine pits, surface mostly smooth and shiny; mesopleurum impunctate centrally, otherwise with mostly moderately dense, fine pits, surface mostly smooth and shiny; 1st and 2nd terga smooth and shiny. *Shape*: Clypeal margin slightly upturned (Fig. 78:B); flagellum moderately slender (Fig. 80:B), slightly longer than wide to apex; areola broadly hexagonal, elongate apically (Fig. 85:B), area dentipara nearly regular, apophysis distinct, propodeal carinae distinct; 1st tergum slender (Fig. 84:G), postpetiole distinctly narrow, 2nd tergum distinctly narrowed; femur III moderately slender (Fig. 79:B), ratio 0.22; radial cell long, Rs straight. *Other*: Wings a little tinted, 5th and 6th abdominal sterna membranous medially, tyloids (2) short (Fig. 87:D), glumes moderately dense (Fig. 88:B).

FEMALE DIAGNOSIS. Large, 7.5 mm long; legs black except tibiae I and II white dorsally; propodeum slightly sloped, carinae weak, area dentipara narrowed, apophysis distinct and obliquely projecting; 3rd valvula long, about as long as hind tarsomeres; face rugulose with very dense, fine pits; clypeus black, margin very slightly upturned (Fig. 78:C); flagellum slender, distinctly narrowed apically (Fig. 80:E), segments 5-9 white; wings distinctly tinted; radial cell long, Rs straight, very slightly bowed.

FEMALE DESCRIPTION (based on one specimen). *White*: Flagellar segments 5-9 and 10 basally; tibiae I and II dorsally. *Orange*: Abdomen dark orange except basally and apically, and 3rd valvula. *Black*: Flagellum except annulus, scape, clypeus, mandible, tegula, legs except tibiae I and II dorsally, and abdomen basally, apically, and 3rd valvula. *Punctation* (Fig. 86:B,E): Face and frons rugulose with very dense, fine pits; temple and propleurum slightly rugulose with variably dense pits; mesopleurum with variably dense to sparse, distinct pits, surface a little rugulose; metapleurum rugulose. *Shape*: Flagellum slender and distinctly narrowed apically, basal 3 segments elongate (Fig. 80:E,I); clypeal margin very slightly upturned (Fig. 78:B); areola broadly hexagonal, area dentipara distinctly narrowed, a little elongate apically, apophysis distinct and obliquely projecting, propodeum a little sloped; 1st tergum moderately slender (Fig. 84:C), postpetiole more or less gradually widened, about twice as wide as long; femur III moderately swollen (Fig. 79:D), ratio about 0.22; radial cell long, Rs straight, very slightly bowed. *Other*: 3rd valvula long, about as long as hind tarsomeres; wings distinctly tinted.

REMARKS. This species differs from all other species of the Santacruzensis Group by the distinct white on the front and middle tibiae dorsally, a feature common in the Texanus Group. It is placed here because the male has a narrowed postpetiole and 2nd tergum, abdominal sterna 5 and 6 are membranous medially, and the clypeal margin is distinctly upturned. The female differs from other Santacruzensis Group species by the white on flagellar segments 5-9 and on the front and middle tibiae.

RANGE. Southwestern United States in Arizona, New Mexico, and western Texas. Collections May, July, and August.

MATERIAL EXAMINED. **Holotype**: male, near Alpine ARIZONA, 29-v-1947, H. and M. Townes (AEI). **Paratypes**: 2 M, 1 F (AEI, UCR): ARIZONA, NEW MEXICO, TEXAS.

ETYMOLOGY. (Greek) *Leuco-* (white) + *cnemis* (tibia), referring to the white on the tibiae.

Endasys occipitis Luhman, sp. n.
(Figs. 2:A-D, 89:B; Map 5)

MALE DIAGNOSIS. Moderately large, 7-7.5 mm long; face rugulose, with very dense, coarse pits; clypeus black, impressed apically, margin distinctly upturned (Fig. 78:D); area dentipara narrowed (Fig. 83:C), apophysis moderately distinct; 1st abdominal tergum slender (Fig. 84:G), postpetiole squarish, widening apically; 2nd tergum wide; legs I and II yellowish orange, leg III mostly orange except tibia black apically and tarsomeres blackish orange; abdomen orange except sometimes basally; flagellum moderately stout, segments on apical half square (Fig. 80:C); vertex slightly lengthened, flattened, and sloped (Fig. 89:B); wings slightly tinted; tibia III dorsally with about 3 rows of spines; Arizona and northern Mexico.

MALE DESCRIPTION. *Yellowish orange*: Scape, tegula, legs I and II, hind trochanters. *Orange*: Leg III except usually tibia apically and hind tarsomeres; abdomen except often basally. *Black*: Clypeus, mandible often orangish black, usually tibia III apically, often abdomen basally. *Punctation* (Fig. 86:E-F): Face rugulose, with very dense, coarse pits; frons slightly rugulose, with variably dense, coarse pits; temple slightly rugulose, with variably dense pits; propleurum sparsely punctate to impunctate, pits variable; mesopleurum impunctate centrally, with very slight rugulosity, remaining areas variably coarse and rugulose; postpetiole and 2nd tergum smooth and shiny. *Shape*: Clypeus slightly impressed apically, with upturned margin; flagellum moderately stout, segments squarish on apical half; areola broadly hexagonal, area dentipara regular, with weak apophysis; 1st tergum slender, postpetiole squarish, widening to apex; 2nd tergum wide; femur swollen (Fig. 79:D), ratio 0.25. *Other*: Vertex slightly lengthened, flattened, and sloped, wings slightly tinted, abdominal sterna 5-6 membranous medially, glumes moderately dense, hind tibial spines in about 3 rows.

FEMALE DIAGNOSIS. Moderately large, 6.6-7 mm long; face and frons rugulose, with very dense, coarse pits; clypeus black, impressed apically, margin upturned (Fig. 78:D); flagellum bicolored, slender, narrowed apically (Fig. 80:E); temple rugulose, with dense, coarse pits; area dentipara narrowed, apophysis moderately distinct; 1st abdominal tergum moderately slender (Fig. 84:C), postpetiole gradually widened; femur III swollen (Fig. 79:E), ratio 0.30; legs I and II yellowish orange, leg III and abdomen orange; 3rd valvula moderately short, about as long as basal 3 hind tarsomeres.

FEMALE DESCRIPTION. *Yellowish orange*: Flagellum basal half, scape, tegula, legs I and II. *Orange*: Mandible except basally, leg III, abdomen. *Black*: Flagellum apical half (dorsally), clypeus, mandible basally. *Punctation* (Fig. 86:E-F): Face and frons rugulose, with very dense, coarse pits; temple rugulose, with variably dense, coarse pits; propleurum slightly rugulose, with mostly sparse, coarse pits; mesopleurum rugulose with variably dense, coarse pits. *Shape*: Clypeus slightly impressed apically, margin upturned; flagellum slender, narrowed apically, segments 1-3 elongate (Fig. 80:E,I); areola broadly hexagonal, area dentipara narrowed, apophysis moderately distinct and somewhat wide; femur III swollen, ratio 0.30; 1st tergum moderately stout to moderately slender, postpetiole gradually widened to about 1.75 times as wide as long. *Other*: Vertex slightly lengthened, impressed, and sloped; 3rd valvula moderately short, about as long as basal 3 hind tarsomeres; wings slightly tinted.

REMARKS. This species is distinctive among others in the Santacruzensis Group by the elongate and flattened vertex. It may be related to *durangensis*, from which males can be distinguished by the yellowish orange coxae I and II, orange coxa III, and slightly widened postpetiole and 2nd tergum; females are distinguished by the narrow flagellum, in addition to the flattened vertex.

RANGE. Southern Arizona and northern Mexico. Collections May and (mostly) August through October.

MATERIAL EXAMINED. **Holotype**: Male, Parker Cyn. L., ARIZONA, 22-viii-1974, H. and M. Townes (AEI). **Paratypes**: 6 M, 2 F (AEI): ARIZONA, NUEVO LEON.

ETYMOLOGY. *Occipitis* (of the occiput), referring to the short occiput.

Endasys melanogaster Luhman, sp. n.
(Fig. 32:A-D; Map 6)

MALE DIAGNOSIS. Medium small, 5-6.5 mm long; 3 tyloids on flagellar segments 10-12; body mostly black except white scape, tegula, coxae I and II apically, and trochanters, remainder of legs I and II yellow; apophysis weak; Arizona.

MALE DESCRIPTION. *White*: Sometimes scape, usually tegula, coxae I and II apically, and trochanters except hind 1st-trochanter. *Yellow*: Usually scape and mandible, sometimes tegula, femora I and II, tibiae I and II, latter pale yellow dorsally, tibia III yellowish black basally. *Black*: Flagellum, clypeus, coxae I and II except apically, coxa III, hind 1st-trochanter, femur III, tibia III more apically, hind tarsomeres, abdomen. *Punctation* (Fig. 86:A-B,E-F): Face rugulose with very dense pits, frons dense with fine pits, temple sparse with very fine pits, propleurum sparse to impunctate with fine pits, mesopleurum impunctate centrally, otherwise very sparsely pitted, and abdominal terga 1 and 2 smooth and shiny. *Shape*: Clypeus impressed apically, margin sharply upturned (Fig. 78:E); flagellum moderately stout (Fig. 80:C), segments square on apical half; areola broadly hexagonal, a little longer apically, area dentipara regular, apophysis weak, propodeal carinae distinct; 1st tergum slender (Fig. 84:G), postpetiole longer than wide, 2nd tergum regular to slightly narrowed; femur III moderately swollen (Fig. 79:C), ratio 0.22-0.23; Rs very slightly curved. *Other*: 3 tyloids on flagellar segments 10-12, basal 2 longer than distal 3rd; wings slightly tinted; 5th abdominal sternum membranous medially.

FEMALE. Unknown.

REMARKS. See discussion under *arizonae*.

MATERIAL EXAMINED. **Holotype**: male, Parker Ck., Sierra Ancha, ARIZONA, 9-v-1947, H. and M. Townes (AEI). **Paratypes**: 10 M (AEI, Dasch): ARIZONA.

ETYMOLOGY. (Greek) *Melano-* (black) + *gaster* (stomach), referring to the black on the abdomen.

Endasys pinidiprionis Luhman, sp. n.
(Fig. 59:A-D; Map 12)

MALE DIAGNOSIS. Large, 7.5 mm long; clypeus blackish yellow, apically depressed with margin very strongly upturned (Fig. 78:E); face and frons rugulose with very dense, coarse pits, propleurum and mesopleurum variably punctate with distinct pits except central area of latter impunctate; area dentipara nearly regular or slightly narrowed, apophysis small but distinct, propodeal carinae distinct; 1st tergum moderately slender (Fig. 84:F), postpetiole squarish; coxa III black with white apically, femur III black, tibia III yellow with black apically, coxae I and II mostly white, all trochanters white; flagellum moderately slender (Fig. 80:B), segments longer than wide to apex; montane Mexico.

MALE DESCRIPTION (based on one specimen). *White*: Scape, tegula, coxae I and II except basally, coxa III apically, trochanters. *Yellow*: Clypeus blackish yellow, mandible, femora I and II, tibiae I and II except some blackish apically, tibia III except apically. *Orange*: Abdomen except basally and apically. *Black*: Coxae I and II basally, coxae III except apically, femur III, tibia III apically, hind tarsomeres, abdomen basally and apically. *Punctation* (Fig. 86:D-F): Face rugulose with very dense, coarse pits; frons with very dense, coarse pits; temple with dense, more or less distinct pits; propleurum with dense to sparse, distinct pits, rugulose lower part; mesopleurum impunctate centrally, otherwise with sparse, distinct pits; 1st and 2nd terga smooth and shiny. *Shape*: Clypeus depressed apically with margin sharp and distinctly upturned (Fig. 78:E); flagellum moderately slender (Fig. 80:B), segments a little longer than wide on apical half; areola broadly hexagonal, area dentipara nearly regular or slightly narrowed, apophysis small, moderately distinct, propodeal carinae distinct; 1st tergum moderately slender (Fig. 84:F), postpetiole squarish, 2nd tergum regular or slightly widened; femur III moderately slender (Fig.

79:B), ratio 0.21; Rs straight, areolet wide. *Other*: Tyloids (2) yellowish, distal one longer than proximal; wings hyaline; 5th abdominal sternum membranous medially.

FEMALE DIAGNOSIS. Large, 8 mm long; face and frons rugulose with very dense, coarse pits, temple and propleurum variably dense with coarse pits, mesopleurum evenly dense with fine pits; cheek distinctly swollen; flagellum black, a little swollen beyond middle, narrowed apically (Fig. 80:E); coxae II and III black, trochanters yellow, femur III black, tibia III and tarsomeres orange, abdomen orange except black basally; femur III moderately swollen (Fig. 79:C), ratio 0.25; 3rd valvula moderately short, as long as basal 3 hind tarsomeres.

FEMALE DESCRIPTION (based on one specimen). *Yellow*: Legs I and II except coxa of latter. *Orange*: Scape blackish orange, mandible, tegula, coxae II and III ventrally, tibia III, hind tarsomeres, abdomen except basally and 3rd valvula. *Black*: Flagellum, clypeus, coxae II and III except ventrally, femur III, and abdomen basally and 3rd valvula. *Punctation* (Fig. 86:B,F): Face and frons rugulose with very dense, coarse pits; temple and propleurum rugulose with variably dense, coarse pits; mesopleurum rugulose with evenly dense pits; metapleurum a little rugulose with dense, mostly fine pits. *Shape*: Clypeus depressed apically, margin distinctly upturned (Fig. 78:E); cheek distinctly swollen; flagellum moderately stout, a little swollen beyond middle, narrowed apically, basal 3 segments a little longer than wide; areola broadly hexagonal, area dentipara narrowed, apophysis distinct, propodeal carinae distinct; 1st tergum moderately stout, postpetiole about twice as wide as long; femur III moderately swollen (Fig. 79:C), ratio 0.25; radial cell short, Rs mostly straight. *Other*: 3rd valvula moderately short, as long as basal 3 hind tarsomeres; wings hyaline.

REMARKS. This species is related to *durangensis* and *flavissimus*. Males differ from *durangensis* by the black hind femur, yellowish hind tibia, and distinctly yellow clypeus, without blackish; and from the *flavissimus* by lacking the overall yellow of the legs and abdomen. Females differ from *durangensis* by the black flagellum, narrowed on apical half, and the black hind coxa and femur; females of *flavissimus* are unknown.

RANGE. Michoacan, Mexico. Reared April and May.

HOST. Diprionidae: *Diprion* sp., on *Salix* (USNM).

MATERIAL EXAMINED. **Holotype**: male, Uruapan, MICHOACAN, iii-1970, S. Cisneros (USNM). **Paratype**: 1 F (USNM): MICHOACAN.

ETYMOLOGY. *Pini-* (pine) + *diprionis* referring to the host of the type series.

Endasys punctatior Luhman, sp. n.
(Fig. 71:A-D; Map 6)

MALE DIAGNOSIS. Medium small, 5-6 mm long; white coxae I and II and coxa III apically, femur III orange with black apically, tibia III orange with black basally and apically, postpetiole and abdominal terga 2-5 mostly orange, remaining terga black, sometimes black apically and laterally on terga 2-5; clypeus black, margin sharply upturned apically (Fig. 78:E); frons with dense, fine, distinct pits; 1st tergum distinctly slender (Fig. 84:G), nearly parallel sided, postpetiole elongate; southern California.

MALE DESCRIPTION. *White*: Usually scape and mandible, tegula, coxae I and II except basally, coxa III apically, and all trochanters. *Yellow*: Sometimes scape and mandible, femora I and II, tibiae I and II. *Orange*: Femur III except apically, tibia III except basally and dorsally, most of abdomen except basally, apically, and apically and laterally on terga 2-5. *Black*: Clypeus, coxae I and II basally, hind coxa III apically, femur III apically, tibia III basally and apically, hind tarsomeres, abdomen basally, apically, and apically and laterally on terga 2-5. *Punctation* (Fig. 86:A-D): Face evenly dense with fine pits; frons dense with fine pits; temple sparse with indistinct pits; propleurum variably dense to sparse with more or less distinct pits; mesopleurum impunctate centrally, otherwise sparse with more or less distinct, fine pits; 1st and 2nd terga smooth and shiny. *Shape*: Clypeus wide, margin sharply upturned (Fig. 78:E); flagellum moderately slender (Fig. 80:B), segments a little longer than wide; areola broadly hexagonal, area dentipara regular, apophysis weak, propodeum slightly elevated, carinae weak; 1st tergum slender (Fig. 84:G), nearly parallel-sided, postpetiole distinctly elongate, 2nd tergum narrowed or regular; femur III moderately swollen (Fig. 79:C), ratio 0.23; Rs straight (Fig. 82:B). *Other*: Sometimes 3rd tyloid on 12th flagellar segment; wings hyaline; 5th abdominal sternum complete medially; glumes dense.

FEMALE DIAGNOSIS. Small, 4-5 mm long; 3rd valvula moderately long, about as long as basal 4 hind tarsomeres, 1st abdominal tergum slender, postpetiole squarish (Fig. 84:C), face and frons very densely and coarsely punctate, flagellum moderately stout and linear (Fig. 80:F), mostly orangish, leg III orange except coxa blackish, and abdomen except basally black.

FEMALE DESCRIPTION. *Orange*: Flagellum, scape, clypeus, mandible, tegula, legs I and II, leg III except most of coxa, abdomen except basally. *Black*: Coxa III mostly blackish, sometimes femur III with a little blackish, and basal 0.75 of 1st tergum, sometimes additional terga blackish laterally and apically. *Punctation* (Fig. 86:B,D,F): Face and frons rugulose with very dense, coarse pits; temple with sparse, more or less distinct pits; propleurum with dense, more or less distinct pits; mesopleurum with variably sparse punctation, very sparse centrally, pits fine, more or less distinct. *Shape*: Clypeal margin sharply upturned (Fig. 78:E); flagellum moderately stout, linear to apex, basal 3 segments moniliform (Fig. 80:F-G); areola nearly hexagonal, a little elongate anteriorly (Fig. 85:A), area dentipara narrowed, apophysis weak, propodeum moderately swollen dorsally, carinae weak; 1st tergum slender (Fig. 84:C), postpetiole square; femur III swollen (Fig. 79:E), ratio about 0.30; radial cell short (Fig. 82:D), about as long as 2nd-discoidal cell, Rs curved. *Other*: 3rd valvula moderately short, about as long as basal 4 hind tarsomeres; wings a little tinted.

REMARKS. See discussion under *santacruzensis*.

RANGE. California. Collections in April.

MATERIAL EXAMINED. **Holotype**: male, Potrero, San Diego Co., CALIFORNIA, 8-iv-1974, H. and M. Townes (AEI). **Paratypes**: 22 M, 22 F (AEI, CAS, UCB, UCR): CALIFORNIA. Other material studied: 3 M, 4 F (CAS, UCB, UCD): CALIFORNIA.

ETYMOLOGY. *Punctat-* (punctate) + *-ior* (comparative suffix), referring to the more deeply punctate frons.

Endasys santacruzensis Luhman, sp. n.
(Fig. 5:A-D; Map 14)

MALE DIAGNOSIS. Medium small, 5-6 mm long; clypeal margin distinctly upturned, abdomen mostly black basally, apically, and laterally and apically on terga 2-6, remaining areas orangish; white scape, mandible, tegula, coxae, and trochanters; blackish clypeus, coxa III except apically, femur III and tibia III; area dentipara widened (Fig. 83:A), with weak carinae and apophysis; flagellum moderately slender, segments longer than wide to apex (Fig. 80:B), glumes sparse; coastal California.

MALE DESCRIPTION. *White*: Scape, mandible, tegula, front and middle coxae except basally, coxa III apically, hind trochanters. *Yellow*: Clypeus sometimes blackish yellow, femora I and II, tibiae I and II. *Orange*: Clypeus usually blackish orange, coxae I and II basally, usually 1st abdominal tergum apically, variably on terga 2-6, femur III orangish basally, sometimes tibia III except apically. *Black*: Flagellum, sometimes clypeus, coxa III except apically, most of femur III, tibia III at least apically, abdomen usually basally and apically, laterally and apically on terga 2-6; darker forms with terga 2-6 mostly black. *Punctation* (Fig. 86:A-D): Face appearing granular, frons with dense, fine pits, temple with sparse, indistinct pits, propleurum with dense to very sparse punctation, more or less distinct pits; mesopleurum impunctate, otherwise with sparse, more or less distinct pits; postpetiole and 2nd tergum smooth and shiny. *Shape*: Clypeal margin distinctly upturned (Fig. 78:E); flagellum moderately slender (Fig. 80:B), segments longer than wide to apex; areola elongate hexagonal; area dentipara widened (Fig. 83:A), apophysis weak or indistinct; femur III moderately swollen (Fig. 79:C), ratio 0.23-0.24. *Other*: Wings hyaline, 5th abdominal sternum complete, sometimes short 3d tyloid on 12th flagellar segment, glumes sparse (Fig. 88:C).

FEMALE DIAGNOSIS. Medium small, 5-6 mm long; clypeus impressed apically with distinctly upturned margin (Fig. 78:D), flagellum slender, narrowed apically (Fig. 80:E), postpetiole parallel-sided, 3rd valvula about as short as basal 2 hind tarsomeres, abdomen blackish basally and apically, legs I and II yellow, coxa III blackish, femur III blackish except basally and apically, tibia III orange, areola broadly hexagonal, area dentipara narrowed with more or less distinct apophysis, radial cell short, Rs very weakly curved.

FEMALE DESCRIPTION. *Yellow*: Tegula, legs I and II, and hind trochanters. *Orange*: Flagellum blackish orange, sometimes clypeus blackish orange at least apically, scape, mandible, femur III except basally and apically, tibia III, variably on abdominal terga 2-6 except often laterally and apically. *Black*: Usually clypeus, coxa III orangish black, femur III orangish black except basally and apically, abdomen basally and apically, and often laterally and apically. *Punctation* (Fig. 86:B,D): Face slightly rugulose or granular; frons with dense, more or less distinct pits; temple with sparse, finely distinct pits; propleurum with dense, more or less distinct pits; mesopleurum with sparse, more or less distinct pits and very slightly rugulose. *Shape*: Flagellum slender and narrowed on apical half, first 3 segments short (Fig. 80:E,H); clypeus impressed apically with margin distinctly upturned (Fig. 78:E); areola elongate hexagonal (Fig. 85:A), area dentipara narrowed, apophysis more or less distinct, propodeal carinae weak; 1st abdominal tergum moderately stout (Fig. 84:B), postpetiole moderately widened, about 1.7 times as wide as long, parallel-sided; femur III swollen (Fig. 79:E), ratio 0.27-0.29. *Other*: 3rd valvula short, as long as basal 2 hind tarsomeres, wings mostly clear, radial cell short, Rs slightly curved.

REMARKS. This species is closely related to *tricoloratus* and *punctatior*. Males of *santacruzensis* are separated from both species by the black clypeus and mostly blackish femur III, tibia III, and abdomen. Females are separated by the distinctly narrowed flagellum, basal 3 segments moniliform, black clypeus, and short 3rd valvula, about as long as basal 2 hind tarsomeres.

RANGE. Mostly coastal California from Contra Costa Co. to San Diego Co. Collections April and May.

MATERIAL EXAMINED. **Holotype**: male, Julian, San Diego Co., CALIFORNIA, 8-v-1974, H. and M. Townes (AEI). **Paratypes**: 126 M, 38 F (AEI, CAS, OSU, UCB): CALIFORNIA. Other material studied (AEI, CAS, CNC, UCB, UCD, UCR, USNM): CALIFORNIA.

ETYMOLOGY. From Santa Cruz Island, California, to note occurrence of species there.

Endasys spinissimus Luhman, sp. n.
(Fig. 76:A-D; Map 1)

MALE DIAGNOSIS. Large, 7.3 mm long; face finely rugulose with very dense pits; clypeus blackish and blackish orange, margin more or less upturned (Fig. 78:C); frons with evenly dense, fine pits; white scape, tegula, coxae I and II apically, trochanters I and II, and tibiae I and II dorsally; coxae I and II black basally, coxa III entirely black, femur III mostly black with orange basally, tibia III mostly black; abdomen orange except basally and parameres; hind tibial spines in 3 distinct rows; propodeal carinae moderately weak, area dentipara slightly widened, apophysis weak and inconspicuous; 1st tergum moderately slender (Fig 84:F), postpetiole squarish; femur III moderately slender (Fig. 79:B), ratio 0.21; Durango, Mexico.

MALE DESCRIPTION (based on one specimen). *White*: Scape, tegula, coxae I and II apically, trochanters except hind 1st-trochanter, tibiae I and II dorsally. *Yellow*: Most of femora I and II and tibiae I and II except latter apically. *Orange*: Clypeus and mandible apically blackish orange, femur III basally, and abdomen except basally and parameres. *Black*: Flagellum, clypeus and mandible basally, coxae I and II basally, coxa III, hind 1st-trochanter, femur III except basally, tibia III except orangish black ventrally, and abdomen basally and parameres. *Punctation* (Fig. 86:C-F): Face finely rugulose with very dense pits; frons slightly rugulose with evenly dense pits; temple with dense, indistinct pits; propleurum with dense to very sparse punctation, more or less distinct pits; mesopleurum impunctate centrally, otherwise with sparse, distinct pits; metapleurum shiny, mostly smooth; 1st and 2nd terga smooth and shiny. *Shape*: Clypeal margin more or less upturned (Fig. 78:C); flagellum moderately slender (Fig. 80:B), segments longer than wide; areola broadly hexagonal, area dentipara slightly wide and widened laterally (Fig. 83:A), apophysis weak and indistinct; 1st tergum moderately slender (Fig. 84:F), postpetiole square, 2nd tergum regular; femur III moderately slender (Fig. 79:B), ratio about 0.21; radial cell wide, greater than width of 2nd-discoidal cell, Rs straight. *Other*: Tyloids (2) moderately long and prominent; wings slightly tinted; 5th abdominal sternum membranous medially; glumes very dense (Fig. 88:D); hind tibial spines in 3 rows.

REMARKS. The type is the only specimen. This species is distinct from other species in Mexico and Arizona by the combined characters in the diagnosis. It may be related to *durangensis*.

HOLOTYPE. Male, 5 km W. El Salto, 8800 ft., DURANGO, 12-vi-1972, B. and C. Dasch (AEI).

ETYMOLOGY. *Spini-* (spiny) + *-issimus* (-est), referring to the distinctly spiny hind tibia.

Endasys tetratylus Luhman, sp. n.
(Fig. 8:A-D; Map 12)

MALE DIAGNOSIS. Medium large, 6-7 mm long; 4 tyloids on flagellar segments 10-13, black clypeus with margin sharply upturned (Fig. 78:E), yellow scape and mandible, area dentipara narrowed with weak apophysis, propodeal carinae moderately strong, wings hyaline, femur III swollen (Fig. 79:D), ratio 0.24; legs I and II mostly yellow with blackish and yellow coxae and 1st trochanters, coxa III mostly black, hind trochanters blackish and yellowish, femur III orangish with apex black, tibia III mostly black, abdomen shiny and orange with black basally and apically, 1st tergum moderately slender (Fig. 84:F), and postpetiole squarish; southern Arizona.

MALE DESCRIPTION. *Yellow*: Scape, mandible, coxae I and II except basally, coxa III apically, trochanters ventrally except 1st-trochanters basally, femora I and II, tibiae I and II. *Orange*: Femur III orangish black basally, abdomen except basally and apically. *Black*: Flagellum, clypeus, tegula more brownish, coxae I and II basally, coxa III except apically, trochanters ventrally, femur III more orangish black basally, tibia III except lighter area sub-basally, hind tarsomeres, and abdomen basally and apically. *Punctation* (Fig. 86:B,D-E): Face rugulose with very dense pits; frons variably dense with finely distinct pits; temple dense with more or less distinct pits; propleurum dense to sparse with finely distinct pits; mesopleurum mostly impunctate centrally and smooth, otherwise sparse with more or less distinct pits; postpetiole and 2nd tergum smooth and shiny. *Shape*: Clypeus sharp and distinctly upturned (Fig. 78:D); flagellum moderately stout (Fig. 80:C), segments squarish on apical half; areola broadly hexagonal, area dentipara narrowed, apophysis very weak, indistinct, porpodeal carinae moderately strong; 1st abdominal tergum moderately slender (Fig. 84:F), postpetiole square, 2nd tergum regular; femur III swollen (Fig. 79:D), ratio about 0.24. *Other*: 4 tyloids on flagellar segments 10-13, distal 2 shorter than basal 2;

wings hyaline, 5th abdominal sternum membranous medially; glumes moderately sparse on flagellum basally, moderately dense apically; hind tibial spines distinct and in 3 rows.

FEMALE DIAGNOSIS. Medium large, 6-7 mm long; clypeus strongly upturned (Fig. 78:E); face and frons rugulose with very dense, coarse pits; mesopleurum rugulose with mostly sparse pits; 1st abdominal tergum moderately slender (Fig. 84:C), postpetiole gradually expanded to about 1.5 times as wide as long; wings hyaline; flagellum orange, moderately slender and narrowed on apical half (Fig. 80:E); legs and abdomen orange; southern Arizona.

FEMALE DESCRIPTION. *Orange*: Flagellum, scape, clypeus orangish black, mandible, tegula, legs, and abdomen. *Punctation* (Fig. 86:B,E-F): Face and frons rugulose with very dense, coarse pits; temple with sparse to dense, coarse pits; propleurum with dense, fine pits, or with very dense, coarse pits; mesopleurum rugulose with mostly sparse pits. *Shape*: Clypeal margin sharply upturned (Fig. 78:E); flagellum moderately slender, narrowed apically (Fig. 80:E), basal 3 segments a little elongate to square; areola broadly hexagonal, area dentipara narrowed, apophysis weak, propodeal carinae moderately strong; 1st tergum moderately slender (Fig. 84:C), postpetiole gradually widened to about 1.5 times as wide as long; femur III swollen (Fig. 79:E), ratio 0.30; Rs mostly straight but slightly curved at apex (Fig. 82:C). *Other*: 3rd valvula moderately short, about as long as basal 3 hind tarsomeres; wings nearly hyaline.

REMARKS. This species appears closely related to *auriger, aureolus,* and *julianus.* Males differ from all of them by the 4 tyloids, yellowish white scape, and coxae blackish with whitish yellow apically. Sometimes *aureolus* has 4 tyloids, but *tetratylus* differs by the whitish yellow on the coxae and trochanters. Female *tetratylus* differs from *auriger* by the flagellum narrow on the apical half, and tibiae I and II without whitish yellow dorsally; and from *aureolus* by the orange flagellum and the short valvula, about as long as basal 3 hind tarsomeres. Females of *julianus* are unknown.

RANGE. Southern Arizona. Collections August and September.

MATERIAL EXAMINED. **Holotype:** male, Portal, ARIZONA, 12-viii-1974, H. and M. Townes (AEI). **Paratypes:** 2 M, 2 F (AEI): ARIZONA.

ETYMOLOGY. (Greek) *Tetra-* (four) + *tylos* (knob), referring to the 4 tyloids on the male antenna.

Endasys tricoloratus Luhman, sp. n.
(Fig. 4:A-D; Map 12)

MALE DIAGNOSIS. Medium small, 5-6 mm long; yellow clypeus with distinctly upturned margin (Fig. 78:D); white scape, coxae I and II, coxa III apically, and trochanters; orange femur III and tibia III except apices, abdomen except basally and apically; 5th abdominal sternum membranous medially; southern California.

MALE DESCRIPTION. *White*: scape, tegula, coxae I and II except basally, coxa III apically, trochanters. *Yellow*: Clypeus except medially usually blackish, mandible, femora I and II, tibiae I and II. *Orange*: Coxae I and II basally, femur III except sometimes apically, tibia III except apically, and abdomen except basally and apically. *Black*: Flagellum, sometimes coxae I and II basally more blackish, coxa III orangish black except apically, sometimes femur III apically, tibia III apically, hind tarsomeres, abdomen basally and apically; sometimes tergum 2 blackish apically. *Punctation* (Fig. 86:A-B,D): Face granular; frons with variably dense, finely distinct pits; temple and propleurum with moderately dense, more or less distinct pits; mesopleurum impunctate centrally, otherwise with variably sparse, more or less distinct pits; postpetiole and 2nd tergum smooth and shiny. *Shape*: Clypeal margin distinctly upturned (Fig. 78:D); flagellum moderately slender (Fig. 80:B), segments longer than wide; areola broadly hexagonal, area dentipara narrowed, apophysis moderately distinct, propodeal carinae weak; 1st abdominal tergum moderately slender (Fig. 84:F), postpetiole longer than wide, 2nd tergum regular; femur III moderately swollen (Fig. 79:C), ratio 0.23-0.24. *Other*: Wings hyaline, areolet small, 5th abdominal sternum membranous medially, and glumes dense.

FEMALE DIAGNOSIS. Small, 4mm long; yellow clypeus with sharp, slightly upturned margin (Fig. 78:C); yellow scape, legs I and II, and hind trochanters; abdomen mostly orange except basally and with some blackish on terga 2-7 laterally and apically; 3rd valvula moderately short, a little longer than the basal 3 hind tarsomeres; postpetiole broadly expanded, sides slightly diverging apically; face rugulose and very densely pitted.

FEMALE DESCRIPTION. *Yellow*: Scape, clypeus, mandible, tegula, legs I and II, hind trochanters. *Orange*: Flagellum, coxa III and femur III more blackish orange, tibia III mostly orange except apically, 1st abdominal tergum apically, and basally on terga 2-7. *Black*: Abdomen basally, terga 2-7 laterally and apically blackish. *Punctation* (Fig. 86:B,D,F): Face and clypeus rugulose and very densely pitted; frons with dense to sparse, more or less distinct pits; temple with sparse, more or less distinct pits; propleurum with moderately dense, finely distinct pits; mesopleurum with mostly sparse, more or less distinct pits. *Shape*: Clypeal margin sharp and slightly upturned (Fig. 78:C); flagellum moderately stout, linear apically, segments 1-3 moniliform (Fig. 80:F-G); areola broadly hexagonal, area dentipara narrowed, apophysis moderately distinct, propodeal carinae weak; 1st abdominal tergum moderately stout (Fig. 84:B), postpetiole square to moderately wide, about 1.7 times as wide as long, side slightly diverging apically; femur III swollen (Fig. 79:E), ratio 0.31; radial cell short, Rs slightly curved. *Other*: 3rd valvula moderately short, about as long as basal 3 hind tarsomeres or longer; wings hyaline.

REMARKS. See discussion under *santacruzensis* and *callidius*.

RANGE. Southern California. Collections April and May.

MATERIAL EXAMINED. **Holotype**: male, Potrero, San Diego Co., CALIFORNIA, 12-iv-1974, H. and M. Townes (AEI). **Paratypes**: 31 M, 1 F (AEI): CALIFORNIA.

ETYMOLOGY. *Tri-* (3) + *coloratus* (colored), referring to the white, yellow, and orange on the head and legs.

Appendix I

Endasys Host List:
Symphyta, Tenthredinoidea (Hymenoptera)

ARGIDAE
A. clavicornis (Fabricius), a birch sawfly: *mucronatus.*
Arge pectoralis (Leach), birch sawfly: *mucronatus.*

DIPRIONIDAE
Diprion similis (Hartig), introduced pine sawfly: *pubescens.*
Diprion sp., on willow: *pinidiprionis.*
Gilpinia hercyniae (Hartig), European spruce sawfly: *patulus.*
Neodiprion abbottii (Leach): *patulus.*
Neod. excitans Rohwer, blackheaded pine sawfly: *patulus.*
Neod. lecontei (Fitch), redheaded pine sawfly: *patulus.*
Neod. merkeli Ross, slash pine sawfly: *patulus.*
Neod. nannulus nannulus Schedl, red pine sawfly: *patulus.*
Neod. pratti banksianae Rohwer, blackheaded jack pine sawfly: *patulus.*
Neod. pratti pratti (Dyar), Virginia pine sawfly: *patulus.*
Neod. sertifer (Geoffroy), European pine sawfly: *patulus, pubescens.*
Neod. swainei Middleton, Swaine jack pine sawfly: *patulus.*
Neod. taedae linearis Ross, loblolly pine sawfly: *arkansensis, patulus.*
Neod. tsugae Middleton, hemlock sawfly: *hesperus.*
Neod. virginianae Rohwer, redheaded jack pine sawfly: *patulus.*
Zadiprion townsendi (Cockerell), bull pine sawfly: *patulus.*

TENTHREDINIDAE
Heterarthrinae
Caliroa cerasi (Linnaeus), pear slug on cherry: *praerotundiceps.*
Metallus rohweri MacGillivray, a *Rubus* leafminer: *praerotundiceps.*
Profenusa sp., a birch leafminer: *praerotundiceps.*
Profenusa sp., a red oak leafminer: *praerotundiceps.*

Nematinae

Nematus currani Ross, a *Populus* sawfly: *nemati.*

Nem. ribesii (Scopoli), imported currantworm: *euryops.*

Nem. salicisodoratus Dyar, a willow sawfly: *brevicornis.*

Nem. tibialis Newman, a willow sawfly: *pubescens.*

(probably) *Nematus* sp., "locust sawfly": *inflatus.*

Pikonema alaskensis (Rohwer), yellowheaded spruce sawfly: *pubescens, callistus.*

Platycampus sp., a *Populus* or *Larix* sawfly: *leioleptus.*

Pristiphora erichsonii (Hartig), larch sawfly: *patulus, pubescens.*

Prist. geniculata (Hartig), mountain ash sawfly: *praerotundiceps.*

Prist. sp., willow sawflies: (probably) *bicolor.*

Appendix II

Paratype Localities of New Species of *Endasys*

albior: COLORADO: Boulder Canyon, Clear Creek Co. (W. Chicago Crk.), Mt. Evans (Doolittle Ranch), Rabbit Ears Pass, Rocky Mt. Nat. Pk. (Poudre L., Phantom Valley), Steamboat Springs.

albitexanus: GEORGIA: Blood Mt.; MICHIGAN: Branch Co.; MINNESOTA: Ramsey Co. (St. Paul); NEW YORK: Ithaca; ONTARIO: Chaffeys Locks; SOUTH CAROLINA: Cleveland.

angularis: MICHIGAN: Ann Arbor; NEW YORK: Ithaca, Long Is., South Haven, Tompkins Co. (McLean Reservoir).

arizonae: ARIZONA: Chiricahua Mts. (Cave Crk. Cyn.), Coronado Nat. Forest (Molino Basin), Sierra Ancha (Parker Crk.)

arkansensis: ARKANSAS: Calhoun Co., Fordyce, Hampton, Warren, Washington Co.; KENTUCKY: Golden Pond; LOUISIANA: Rapides Parish; MARYLAND: Hardy Co. (Lost River St. Pk.), Montgomery Co. (Colesville), Wheaton; OHIO: Jefferson St. Pk.

aurantifex: ALBERTA: Elkwater; CALIFORNIA: Crescent City, San Gregorio, nr. Sonora Pass (8000 ft.); COLORADO: nr. Estes, Lyon; IDAHO: Idaho City, Lowman, nr. Stanley; MINNESOTA: Clay Co. (Moorhead), Pipestone Co. (Pipestone Nat. Mon.); MONTANA: Bozeman, Glacier Nat. Pk.; NEBRASKA: Valentine; OREGON: Cape Lookout St. Pk., Corvallis, Forest Grove, Hyatt Reservoir. Mt. Hood (5400 ft.), Selma, Takhenitch.

aurarius: CONNECTICUT: Voluntown; GEORGIA: Forsyth; KENTUCKY: Golden Pond; MICHIGAN: Ann Arbor; OHIO: New Concord; ONTARIO: St. Lawrence Is. Nat. Pk. (McDonald Is.), Stittsville; SOUTH CAROLINA: Beaufort Co. (Parris Is.), Cleveland, McClellanville.

aureolus: ARIZONA: Portal.

aurigena: ARKANSAS: Lake Ouachita; FLORIDA: Osceola Nat. Forest; KANSAS: Lawrence; MARYLAND: Takoma Pk.; MICHIGAN: Ann Arbor, Crystal Falls, Pincney Rec. Area (South L.); NEBRASKA: Valentine Refuge;.NEW BRUNSWICK: Kouchibouguac Nat. Pk.; NEW HAMPSHIRE: Base Station; NEW JERSEY: High Point; NEW YORK: Essex Co. (Artist's Br.), Hancock, Ithaca, Onondaga Co.; NORTH CAROLINA: Highlands, Wake; OHIO: Jefferson St. Pk., New Concord, Otsego (McAllister Bio. Sta.), Steubenville, Willis Crk. Res.; ONTARIO: Atikokan; PENNSYLVANIA: Harrisburg; SASKATCHEWAN: Cut Knife (Attons L.); WASHINGTON, D.C. (10 miles E.); WEST VIRGINIA: Bowden, Cranberry Glens, Dolly Soda Wilderness Area, Spruce Knob (4862 ft.).

auriger: ARIZONA: Santa Catalina Mts. (Mt. Lemon, 7800 ft.), Nogales, Parker Canyon L.

bicolorescens: ALBERTA: Waterton; BRITISH COLUMBIA: Atlin; NEW HAMPSHIRE: Mt. Washington (Bigelow Lawn, head Tuck Ravine, Lake of the Clouds); QUEBEC: Great Whale R.; YUKON TERRITORY: Dawson City.

brachyceratus: ALASKA: Anchorage, Mt. McKinley (2500 ft.), Tsaina R.; BRITISH COLUMBIA: Stone Mt. Pk. (2400-3800 ft.); YUKON TERRITORY: Dawson City.

brevicornis: CONNECTICUT: Hamden; ILLINOIS: Union Co., Shawnee St. Forest (Pine Hill); MAINE: Mt. Desert Is. (Northeast Harbor); MARYLAND: Takoma Pk.; MICHIGAN: Ann Arbor, Huntington Woods, Iron Co. (Pentagon Pk.), Metro Pk. (Delhi-Huron, Hudson Mills); MINNESOTA: Aitken, Big Fork, Clay Co. (Buffalo St. Pk.), Houston Co. (Beaver Crk. Valley St. Pk.), Itasca Co. (nr. Grand Rapids), Itasca Pk., Pine Co. (Cloverdale), Pipestone Co. (Pipestone Nat. Mon.), Polk Co. (nr. Crookston, Red Lake Reservoir), Ramsey Co. (oak forest), Split Crk. St. Pk.; MISSOURI: Williamsville; NEW BRUNSWICK: Kouchibouguac Nat. Pk.; NEW JERSEY: Moorestown; NEW YORK: Bemus Point, Canajoharie, Farmingdale, Hancock, Ithaca, Tompson Co. (McLean Reservoir); NORTH CAROLINA: Highlands; OHIO: Findlay St. Pk., New Concord, Otsego (McAllister Bio. Sta.); ONTARIO: Algonquin Pk., Guelph. Pt. Pelee, St. Lawrence Is. Nat. Pk. (McDonald Is., Thwartway Is.), Vineland; QUEBEC: Quebec City (Laval Univ.); SOUTH CAROLINA: Cleveland, nr. Tigerville, Pickens Co. (Wattacoo); TENNESSEE: White House; WEST VIRGINIA: Spruce Knob (4862 ft.).

callidius: ARIZONA: Cochise Co. (Portal, S.W. Res. Sta.), Oak Creek Canyon; CALIFORNIA: San Diego Co. (Potrero).

callistus: MAINE: Howland, Mt. Katahdin; MICHIGAN: Ann Arbor; MINNESOTA: Aitken Co., Big Fork, Itasca Co. (nr. Grand Rapids), Lyon Co. (Camden St. Pk.), Pope Co. (Glacial Lakes St. Pk.); NEBRASKA: Valentine Refuge; NEW YORK: Ithaca; OHIO: Jefferson St. Pk., New Concord; PENNSYLVANIA: Gaines; WEST VIRGINIA: Bowden.

chiricahuanus: ARIZONA: nr. Alpine, Cochise Co. (nr. Portal, S.W. Res. Sta., 5400 ft.), Chiricahua Mts. (Cave Creek Cyn.); NEW MEXICO: Cimarron (8000-10,000 ft.).

chrysoleptus: MARYLAND: Takoma Pk.; MICHIGAN: Huron Mts.; NEW YORK: Ithaca; NORTH CAROLINA: Buncombe Co. (Craggy Gardens, 5400 ft.), Clingman Dome (6600 ft.), Hamrick, Mt. Mitchell (4000-6000 ft.), Pisgah Mt. (4600-5749 ft.), Wilkes Co. (Scenic Hwy.), Yancey Co. (Crabtree Meadows, 3600 ft.); WEST VIRGINIA: Spruce Knob.

concavus: ARIZONA: Chiricahua Mts. (Cave Creek Cyn.), Sierra Ancha (Parker Crk.), Huachuca Mts. (Bear Cyn.).

coriaceus: ALASKA: Thompson Pass; BRITISH COLUMBIA: Stone Mt. Pk. (3500-3800 ft.); SASKATCHEWAN: Waskesiu.

daschi: COLORADO: Gould, Morley, Mt. Evans (Doolittle Ranch, 9800 ft.), Rabbit Ears Pass (9500 ft.), Rocky Mt. Nat. Pk. (Phantom Valley, Poudre L.), Steamboat Spring, West Chicago Crk. (9800 ft.); IDAHO: nr. Stanley (Galena Summit); MONTANA: Glacier Nat. Pk.

declivis: BRITISH COLUMBIA: Stone Mt. Pk. (3800-5500 ft.).

durangensis: DURANGO: nr. El Salto (San Antonio); MEXICO, D.F.: Temescaltepec (Real de Arriba).

elegantulus: BRITISH COLUMBIA: nr. Terrace (Mt. Thornhill); KANSAS: Riley Co.; MICHIGAN: Ann Arbor, Brighton R.A. (Bishop L.), Crystal Falls; OHIO: New Concord (Oak Park), Otsego (McAllister Bio. Sta.), Willis Crk. Reservoir; SOUTH CAROLINA: Columbia.

euryops: ALASKA: Alcan Hwy. (Gardner Crk. Camp); BRITISH COLUMBIA: Burns L.; MICHIGAN: Ann Arbor; MINNESOTA: Cook Co. (Hovland), Itasca Co. (nr. Grand Rapids), Pope Co. (Glacial Lakes St. Pk.); NEW BRUNSWICK: Kouchibouguac Nat. Pk.; NEWFOUNDLAND: Labrador (Cartwright); NEW YORK: Ithaca, Long Island (Dix Hills), Oswego; OREGON: Corvallis; WASHINGTON: Mt. Rainier (5500 ft.); YUKON TERRITORY: Rampart House.

flavivittatus: CALIFORNIA: Alameda Co. (cyn. nr. Oakland Naval Hosp.), Berkeley Hills (Strawberry Cyn.), nr. Descanso, Ft. Seward, Julian, Lake Wohlford, Madera Co. (Bass L.), Marin Co. (Fairfax, Mill Valley), Oakland, Potrero, San Jacinto Mts. (Ribbon Wood), Santa Clara Co. (nr. New Almaden, Herbert Crk.).

gracilis: CALIFORNIA: Camino, Julian, Lake Wohlford, Mendocino Co. (Hopland Field Sta.), San Diego Co. (Potrero), Santa Barbara Co. (Santa Cruz Is., Field Sta.), Santa Clara Co. (New Almaden, Herbert Crk.).

granulifacies: CALIFORNIA: Alpine Co. (Hope Valley), Contra Costa Co. (Marsh Creek Springs), Dardanelle, Lake Tahoe, Mono Co. (Sardine Crk.), Napa Co. (Pope Valley), Oakland, Plumas Co. (nr. Quincy, Buck's L.), Santa Clara Co. (Mt. Hamilton), Tuolumne Co.; OREGON: Corvallis, Hyatt Reservoir, Pinehurst, Selma; WASHINGTON: Mt. St. Helens.

hesperus: BRITISH COLUMBIA: Squamish (Diamond Head Trail, 3800 ft.); CALIFORNIA: Alameda Co. (Berkeley Hills), Camino, Carmel, Cisco, Crescent City, Dardanelle, Donner Pass, El Dorado Co. (Echo L., Fallen Leaf L.), Fish Camp, Humboldt Co. (Arcata), Inverness, Leevining, Mendocino Co. (nr. Mendocino), Mill Valley, Mt. Tallac, Sierra Co. (Gold L., Webber L.), Trinity Co. (Butler Crk. nr. Hyampom, Mt. Meadow Ranch); OREGON: Cannon Beach, Corvallis, Glenada, Lane Co., Hyatt Reservoir, Lake Co. (Chandler St. Pk.), Lane Co. (nr. Noti Bog), Lincoln Co. (Yachats), Mt. Hood (3500 ft.), Pinehurst, Rhododendron, nr. Scio, Seaside, Selma, Silver Falls, Sparks Lake; WASHINGTON: Ashford, Bay Center, Olympia, Mt. Rainier (2700-5300 ft.), Seattle.

hexamerus: ALBERTA: Pincher; BRITISH COLUMBIA: Nanaimo Bio. Sta.; CALIFORNIA: Dardanelle, Devil's Basin (8200 ft.), Fish Camp, Leevining, Nevada Co. (nr. Hobart Mills, Saghen Crk.), Yosemite Pk. (Crane Flat, May L., Tamarack Flat); OREGON: Corvallis, Hyatt Reservoir, Mary's Peak (4000 ft.), Mt. Hood (3500-5400 ft.); SASKATCHEWAN: Waskesiu; WASHINGTON: Ashford, Friday Harbor, Mt. Rainier (2900-5500 ft.), Skagit Co. (Anacortes); WYOMING: Teton Nat. Forest (Snake R.).

julianus: CALIFORNIA: Julian, San Diego Co.

latissimus: ALASKA: Delta Jct., Mt. McKinley Nat. Pk.; ALBERTA: Banff (Mosquito L.); BRITISH COLUMBIA: Atlin, Ft. Nelson, Stone Mt. Pk.; COLORADO: Mt. Evans (Doolittle Ranch); NEWFOUNDLAND: Raleigh.

leioleptus: CALIFORNIA: Carson Pass, Cisco, Donner Pass, Yosemite Pk. (Snow Flat, 8700 ft.); COLORADO: Mt. Evans (Doolittle Ranch, 9800 ft.), Rocky Mt. Nat. Pk. (Fall River Pass, 11,600 ft.); IDAHO: Coeur d'Alene, Craters of the Moon Nat. Mon. (Little Cottonwood Crk.), Lowman, nr. Stanley; OREGON: Corvallis, Crater Lake Nat. Pk. (17,000 ft.), Deschutes Nat. Forest (Smiling R. Camp), Mary's Peak (4000 ft.), Meacham, Mt. Hood (3500-5400 ft.), Ochoco Crk.; WASHINGTON: Bingen, Coleville Nat. Forest (nr. Republic), Mt. Rainier (4000-5000 ft., Yakima Pk.).

leopardus: ALASKA: Anchorage, Delta Junction, Kenai Peninsula (Bertha Crk. Camp), Thompson Pass, Tsaina R.; ALBERTA: Alcan Hwy. (Burwash Flats); BRITISH COLUMBIA:

Racing River (2400 ft.), Stone Mt. Pk. (3800-5500 ft.), nr. Terrace (Lakelse L. Bog); NEWFOUNDLAND: South Branch; YUKON TERRITORY: Alcan Hwy. (Teslin L., Burwash Flats).

leptotexanus: ALASKA: Tsaina R.

leucocnemis: ARIZONA: nr. Alpine, Graham Mts. (Shannoin Camp); NEW MEXICO: Taos Co. (Hondo Cyn.); TEXAS: Guadalupe Mts. (7750 ft.).

melanogaster: ARIZONA: Coronado Nat. Forest (Molino Basin), Sierra Ancha (Parker Crk.).

michiganensis: MAINE: Bar Harbor, Casco; MASSACHUSETTS: Mt. Greylock; MICHIGAN: Crystal Falls, Huron Mts., Iron Co., Mackinaw City, Marquette Co. (Yellow Dog Plains); MINNESOTA: Big Fork; NEW BRUNSWICK: Kouchibouguac Nat. Pk.; NEWFOUNDLAND: South Branch; NEW JERSEY: High Point St. Pk.

nemati: ALASKA: Thompson Pass; ALBERTA: Elkwater Pk.; BRITISH COLUMBIA: Taft.; CALIFORNIA: Crescent City; IDAHO: Lowman, nr. Stanley; OREGON: Cannon Beach, Sparks Lk; WASHINGTON: Ashford, Mt. Rainier (5500 ft.); WYOMING: Shoshone Canyon (6500 ft.); YUKON TERRITORY: Alcan Hwy. (mile 824, Teslin L.).

nigrans: CALIFORNIA: Camino, Dardanelle, Davis, Donner Pass, Monterey Co. (E. of King City), San Joaquin Co. (Tracy), Stanford University; COLORADO: Gould; OREGON: Hyatt Reservoir, Jackson Co. (Ashland Mt.), Mary's Peak (4000 ft.), Mt. Hood (3500 ft.), Ochoco Crk., Selma, Sparks Lake, Three Creeks L., Willamette Nat. Forest (Tombstone Prairie); WASHINGTON: Mt. Rainier (5500 ft.).

obscurus: ALASKA: Anchorage, G. Parks Hwy. (Hurricane Gulch), Thompson Pass, Tsaina R., Turnagain Pass; BRITISH COLUMBIA: Stone Mt. Pk. (3500-3800 ft.); OREGON: Three Creeks L.; YUKON TERRITORY: Alcan Hwy. (Teslin L.), Dawson.

occipitis: ARIZONA: Parker Canyon L., Portal; NUEVO LEON: nr. Linares (San Pedro Iturbide).

oregonianus: BRITISH COLUMBIA: Burnaby Mts., Kokanee Mt., Lac la Hache, So. Fork Crk.; CALIFORNIA: Crescent City, San Mateo Co. (San Mateo Mem. Pk.), Sierra Co. (Gold L.); IDAHO: Coeur d'Alene, Lowman, Priest L., nr. Stanley; OREGON: Corvallis, Mt. Hood, Seaside; WASHINGTON: Mt. Rainier.

pentacrocus: ALASKA: Anchorage, Bonanza Crk., Delta Junction, Thompson; BRITISH COLUMBIA: Racing River (2400 ft.), Stone Mt. Pk. (3800-5500 ft.); COLORADO: nr. Estes;

IDAHO: Lowman (4000 ft.), nr. Stanley; OREGON: Jackson Co. (Ashland Mt.), Mt. Hood Nat. Forest (Breitenbush L.); YUKON TERRITORY: Alcan Hwy. (Burwash Flats).

pinidiprionis: MICHOACAN: Uruapan.

praerotundiceps: CONNECTICUT: Voluntown; IOWA: Ames; KANSAS: Lawrence, Mammoth Cave Pk.; MAINE: Augusta, Canton, Casco, Gilead, Tumbledown Mt. (nr. Weld), Whitefield; MARYLAND: Takoma Pk.; MASSACHUSETTS: Greylock; MICHIGAN: Ann Arbor, Huntington Woods, Manistique, nr. Nahma, Pinckney Rec. Area; NEW BRUNSWICK: Kouchibouguac Nat. Pk.; NEW JERSEY: High Point St. Pk., Ramsey; NEW YORK: Bemus Point, Farmingdale, Ithaca, McLean Bogs, Pulaski, Syracuse; NORTH CAROLINA: Cedar Mt., Highlands, Pisgah Mt. (4800-5300 ft.), Wake Co., Yancey Co. (Crabtree Meadows, 3600 ft.); NOVA SCOTIA: Lun Co. (Bridgewater); OHIO: Findlay St. Pk., Jefferson St. Pk., New Concord, Steubenville; ONTARIO: Chaffey Locks, St. Lawrence Is. Nat. Pk. (Grenadier Is., Thwartway Is., Aubrey Is., McDonald Is.), Sauble Beach, Stittsville; PENNSYLVANIA: Gaines, Spring Branch; QUEBEC: McGregor, St. Etienne; RHODE ISLAND: Westerly; SOUTH CAROLINA: Cleveland; VERMONT: Burlington; Lowell; VIRGINIA: Galax, Skyline Dr.; WEST VIRGINIA: Bolivar, Bowden, spruce Knob (4862 ft.); WISCONSIN: Merrill.

pseudocallistus: MAINE: Southwest Harbor; MICHIGAN: Ann Arbor; MINNESOTA: Houston Co.; NEW YORK: Ithaca; PENNSYLVANIA: Bald Eagle.

punctatior: CALIFORNIA: Atascadero, Humboldt Co. (Arcata), Julian, Lake Wohlford, Mendocino Co. (Hopland Field Sta.), Sacramento Co. (Elk Grove), San Bernardino (Highland), San Diego Co. (Potrero), Santa Cruz Co. (nr. Santa Cruz, Empire Grade Rd.), Santa Cruz Island (U.C. Field Sta.), Stanford University, Tahoe (Alpine Crk.)

rhyssotexanus: ALASKA: Anchorage, Turnagain Pass; BRITISH COLUMBIA: Stone Mt. Pk.; CALIFORNIA: Dardanelle; COLORADO: Rocky Mt. Nat. Pk. (Phantom Valley); OREGON: Mt. Hood.

rubescens: BAJA CALIFORNIA NORTE: nr. Santo Domingo (Hamilton Ranch); CALIFORNIA: Lake Wohlford, Los Cerritos, Nevada Co. (nr. Hobart Mills, Saghen Crk.), Oakland, San Diego Co. (Potrero), Solano Co., Gordon Valley; IDAHO: Lowman (4000 ft.); OREGON: Corvallis, Pinehurst.

rugiceps: COLORADO: Boulder (5500 ft.), Lyons; MICHIGAN: Crystal Falls; MINNESOTA: Itasca Pk., Ramsey Co. (Univ. Minn. Campus); NEW JERSEY: Moorestown; OHIO: Delaware Co., New Concord; ONTARIO: Chaffey Locks, Ojibway; SOUTH CAROLINA: Anderson; TEXAS: Comal Co.

rugitexanus: ALASKA: Bonanza Crk.; MAINE: Mt. Desert; MARYLAND:Plummers Is.; MICHIGAN: Ann Arbor, Crystal Falls; MINNESOTA: Cook Co. (Hovland), Itasca Co. (nr. Grand Rapids). Lyon Co. (Camden St. Pk.), Pipestone Co. (Pipestone Nat. Mon.); NEW BRUNSWICK: Kouchibouguac Nat. Pk.; NEW YORK: Bemus Pt., Ulster Co. (nr. Cherrytown); NORTH DAKOTA: Towers Canyon; OHIO: Jefferson St. Pk., nr. New Concord (Oak Pk.), Otsego (McAllister Bio. Sta.); ONTARIO: Guelph, Ojibway; PENNSYLVANIA: Gaines; WASHINGTON, D.C.; WEST VIRGINIA: Bowden; WISCONSIN: Milwaukee; VERMONT: nr. Jacksonville (Laurel L.); VIRGINIA: Fairfax Co. (Dead Run).

rugosus: BRITISH COLUMBIA: Terrace (Lakelse Bog); MANITOBA: Reynolds; MARYLAND: Wheaton; MICHIGAN: Ann Arbor; NEW BRUNSWICK: Kouchibouguac Nat. Pk., Simcoe; NEW JERSEY: High Point St. Pk.; NEW YORK: So. Andirondacks; OREGON: Pinehurst; PENNSYLVANIA: Spring Branch.

santacruzensis: CALIFORNIA: Contra Costa Co. (Antioch, Marsh Creek Springs), nr. Descanso, Glendale, Lake Wohlford, Mendocino Co., (nr. Branscomb), Monterey Co. (Los Padres Nat. Forest), Oakland, Potrero, San Diego Co. (Ball Jr. College), San Luis Obispo Co. (Cal. Poly. S.U. Campus), Santa Clara Co. (nr. New Almaden, Herbert Crk.), Solano Co. (Green Valley), nr. Sunol.

serratus: BRITISH COLUMBIA: nr. Osoyoos (Richter Pass Rd.), Robson, Victoria; CALIFORNIA: Berkeley, Camino, N. of Leggett; IDAHO: Idaho City, Lava Hot Springs, nr. Stanley (Galena Summit, 8700 ft.); NEVADA: Tuscarora; OREGON: Mary's Peak (4000 ft.), Pinehurst; WASHINGTON: Bingen, Dallesport.

spicus: ALABAMA: Gulf Shores; FLORIDA: Lake Placid; NEW YORK: Long Island (Farmingdale); NORTH CAROLINA: Kill Devil Hills.

taiganus: ALASKA: Anchorage, Bonanza Crk, Skagway (Dewey Glacier, 6000 ft.), Tsaina R.; BRITISH COLUMBIA: Racing River (2400 ft.), Stone Mt. Pk. (3500-3200 ft.); COLORADO: Rocky Mt. Nat. Pk. (Phantom Valley, 9400 ft.; Fall R., 8600 ft.); NEWFOUNDLAND: Raleigh; SOUTH DAKOTA: Spearfish; YUKON TERRITORY: Dawson.

tetratylus: ARIZONA: Portal.

tricoloratus: CALIFORNIA: Julian, Lake Wohlford, San Diego Co. (Potrero).

tyloidiphorus: ALABAMA: Gulfshores; ARKANSAS: Lake Ouachita, Village Creek St. Pk.; ILLINOIS: Huntington; IOWA: Co. 18; KANSAS: Lawrence; KENTUCKY: Golden Pond; MARYLAND: Bowie, Cabin John; MICHIGAN: Ann Arbor, Metro Pk. (Delhi-Huron); MISSOURI: Williamsville; NEW JERSEY: Chatham, High Point St. Pk.; NEW YORK: Bemus Point, Gilbertsville, Ithaca, Long Island (Babylon, Farmingdale, Southhaven), Oneonta, Slaterville, Ulster Co. (nr. Cherrytown); NORTH CAROLINA: Elizabeth, Hamrick,

Highlands, Pisgah Mt. (4800-5300 ft.), Wake Co.; OHIO: Findlay St. Pk., Jefferson St. Pk., New Concord, Otsego (McAllister Bio. Sta.), Steubenville, Willis Crk. Res.; ONTARIO: Ottawa, Walpole Is; PENNSYLVANIA: Bald Eagle, Gaines; RHODE ISLAND: Westerly; SOUTH CAROLINA:. Cleveland, Greenville, McClellanville, Pickens Co. (Wattacoo); TENNESSEE: White House; TEXAS (no locale); WEST VIRGINIA: Bickle Knob (4020 ft.), Bowden.

xanthopyrrhus: ALASKA: Anchorage; ALBERTA: Sturgeon L.; BRITISH COLUMBIA: Stone Mt. Pk. (3800 ft.); OREGON: Mt. Hood (3500-4500 ft.), Sparks L., Steen Mt. (7200 ft.); WASHINGTON: Ashford; YUKON TERITORY: Alcan Hwy. (mile 1105, Burwash Flats).

xanthostomus: ALBERTA: Edmonton; OREGON: Mt. Hood, Ochoco Crk., Pinehurst; IDAHO: Idaho City, nr. Stanley.

References

Ashmead, W.H. 1900. Ann. Rept. New Jersey Bd. Agric. 27 (suppl.): 568.

Barron, J.R. 1975. Provancher's collections of insects, particularly those of Hymenoptera, and a study of the types of his species of Ichneumonidae. Nat. Canad. (extract) 102(4): 387-591.

Beach. 1892. Iowa Agric. Coll Dept. Entomol. Bull.

Benjamin, D.M. 1955. The biology and ecology of the red-headed pine sawfly. USDA Tech. Bull. 1118.

Bentley, G.M. 1940. Insect pest survey. U.S.D.A., Bur. Entomol. and Plant Quar. 20: 191.

Bobb, M.L. 1963. Insect parasites of the Virginia pine sawfly, *Neodiprion pratti pratti*. J. Econ. Entomol. 56(5): 618-621.

Bradley, J.C. 1903. The genus *Platylabus,* Wesmael, with descriptions of 2 new species. Canad. Entomol. 34: 275-283.

Brimley, C.S. 1938. The Insects of North Carolina. N.C. Dept. Agric. 560 pp.

———. 1942. Supplement to insects of North Carolina. N.C. Dept. Agric. 39 pp.

Brischke, C.B. 1891. Schr. Naturf. Ges. Danzig. N.F. vol. 7 (pt. 3): 69.

Brown, A.W. 1941. Foliage insects of spruce in Canada. Canad. Dept. Agric. Tech. Bull. 31: 3-29.

Brues, C.T. 1908. North American Parasitic Hymenoptera, VI. Wisconsin Nat. Hist. Soc. 6(1-2): 50-51.

Carlson, R.W. 1979. Family Ichneumonidae. *In* K.V. Krombein, P.D. Hurd, Jr., et al., Catalog of Hymenoptera in America north of Mexico, Washington, D.C.: Smithsonian Institution Press. vol. 1: 315-740.

Coppel, H.C. 1954. Notes on the parasites of *Neodiprion nanulus* Schedl. Canad. Entomol. 86: 167-168.

Craighead, F.C. 1950. Insect enemies of eastern forests. USDA Misc. Pub. 657.

Cresson, E.T. 1864. Descriptions of North American Hymenoptera in the collection of the Entomological Society of Philadelphia. Phila. Entomol. Soc. Proc. 3: 309-311.

———. 1865. Catalogue of Hymenoptera in the collection of the Entomological Society of Philadelphia from Colorado Territory. Phila. Entomol. Soc. Proc. 4: 265.

———. 1872. Hymenoptera Texana Amer. Entomol. Soc. Trans. 4: 160.

Cushman, R.A. 1922. On the Ashmead manuscript species of Ichneumonidae of Mrs. Slosson's Mt. Washington lists. U.S. Natnl. Mus. Proc. 61: 1-30.

———. 1925. Some generic transfers and synonymy in Ichneumonidae (Hym.). J. Washington Acad. Sci. 15: 388-389.

———. (1926) 1928. Family Ichneumonidae, pp. 920-960. *In* M.D. Leonard (ed.), A list of the insects of New York. Mem. Cornell Univ. Agric. Exp. Sta. 101.

——— and A.B. Gahan. 1921. The Thomas Say species of Ichneumonidae. Entomol. Soc. Washington Proc. 23(7): 153-172.

Dalla Torre, C.G. 1902. *Cryptus,* pp. 558-595, and *Phygadeuon,* pp. 679-697. *In* Catalogus Hymenopterorum, vol. 3: Trigonalidae, Megalyridae, Stephanidae, Ichneumonidae, Agriotypidae, Evaniidae, Pelecinidae.

Davis, G.C. 1898. A review of the Ichneumonidae subfamily Tryphoninae. Amer. Entomol. Soc. Trans. 24: 193-348.

Drooz, A.T., R.C. Wilkinson, and V.H. Fedde. 1977. Larval and cocoon parasites of 3 *Neodiprion* (Hymenoptera, Diprionidae) sawflies in Florida. Env. Entomol. 6(1): 60-62.

Evans, J.D. 1896. List of Hymenoptera taken at Sudbury, Ontario. Canad. Entomol. 28: 9-13.

Finlayson, L.R., and T. Finlayson. 1958. Parasitism of the European pine sawfly, *Neodiprion sertifer* (Geoff.) (Hymenoptera: Diprionidae), in southwestern Ontario. Canad. Entomol. 90: 223-225.

Finlayson, T. 1960. Taxonomy of cocoons and pupae, and their contents, of Canadian parasites of *Neodiprion sertifer* (Geoff.) (Hymenoptera: Diprionidae). Canad. Entomol. 92: 20-47.

———. 1963. Taxonomy of cocoons and puparia, and their contents, of Canadian parasites of some native Diprionidae (Hymenoptera). Canad. Entomol. 95: 475-507.

Fitton, M.G. 1982. A catalogue and reclassification of the Ichneumonidae (Hymenoptera) described by C.G. Thomson. Bull. Brit. Museum (Natural Hist.). 45(1): 119 pp.

Foerster, A. 1868. Synopsis der Familien und Gattung der Ichneumonen. Ver. Naturh. Ver. Preuss. Rheinlande, Verh. 25: 135-227.

Forbes, R.S., G.R. Underwood, F.G. Cuming, and D.C. Eidt. (1959) 1960. Forest insect survey, Maritime Provinces. Canad. Dept. Agric. Ann. Rept. For. Ins. Dis. Survey 1959: 17-31.

Furniss, R.L., and P.B. Dowden. 1941. Western hemlock sawfly *Neodiprion tsugae* Middleton, and its parasites in Oregon. J. Econ. Entomol. 34: 46-52.

Gahan, A.B., and S.A. Rohwer. 1918. Lectotypes of the species of Hymenoptera (except Apoidea) described by Abbe Provancher. Canad. Entomol. 50: 133-137, 166-171.

Gauld, I.D. and G.A. Holloway. 1983. A new genus of Endaseine Ichneumonidae from Australia (Hymenoptera). Contrib. Amer. Entomol. Inst. 20: 191-195.

Girth, H.B., and E.E. McCoy. 1946. Five Ichneumonidae reared from cocoons of the European pine sawfly *Neodiprion sertifer* (Geoff.). J. New York Entomol. Soc. 54: 320.

Gobeil, A. R. 1937. Observations sur la mouche à scie. Nat. Canad. 64: 81-88.

Gravenhorst, I.L.C. 1829. Ichneumonologia Europaea. Vratislavia, pts. 1 and 2.

Gray, J.E. 1845. Cat. Lizards Coll. Brit. Mus. 58 (rept.).

Griffiths, K.J. 1959. Observations on the European sawfly, *Neodiprion sertifer* (Geoff.), and its parasites in southern Ontario. Canad. Entomol. 91: 501-512.

————. 1960. Parasites of *Neodiprion pratti banksianae* Rohwer in northern Ontario. Canad. Entomol. 92: 653-658.

Habermehl, H. 1912. Revision der Cryptiden-Gattung *Stylocryptus* C.G. Thomson unter Berücksichtigung Gravenhorsterschen und Thomsonscher Typen (Hym.). Deutsch. Entomol. Zeitschr. pp. 165-190.

————. 1916. Superrevision der Cryptiden-Gattung *Stylocryptus*. Deutsch. Entomol. Zeitschr. 29: 376-382.

Harris, T.W. 1835. Rept. Geol. Min. Bot. Zool. Mass., 2nd ed., p. 584.

Hendricksen, K.L., and W. Lundbeck. 1918. Landarthropoder (Insecta and Arachnida) conspectus faunae groenlandicae, pt. 2. Medd. om Groenland 22(2): 481-821.

Hetrick, L.A. 1941. Life history studies of *Neodiprion americanum* (Leach). J. Econ. Entomol. 34: 373-377.

Houseweart, M.W., and H.M. Kulman. 1976. Life tables of the yellowheaded spruce sawfly, *Pikonema alaskensis* (Rohwer) (Hymenoptera: Tenthredinidae) in Minnesota. Env. Entomol. 5: 859-867.

Iwata, K. 1960. The comparative anatomy of the ovary in Hymenoptera, pt. V. Ichneumonidae. Acta Hymenopterologica 1(2): 115-169.

Johnson, C.W. 1927. Insect fauna of Nantucket. Biol. Surv. Mt. Desert Region 1: 1-159.

————. 1930. Pub. Nantucket Maria Mitchell Assn. 3(2): 99.

Jussila, R. 1973. Ichneumonidae from Hardangervidda. Fauna Hardangervidda no. 2: 19.

Leconte, J.L. 1859. The complete writings of Thomas Say on the Entomology of North America, vol. 2: 693-694.

Luhman, J.C. 1986. Revision of the Nearctic *Glyphicnemis* Foerster (Hymenoptera: Ichneumonidae, Gelinae). Insecta Mundi 1(3): 133-142.

————— and J. Sawoniewicz. 1991 (in press). Revisions of European species of the subtribe Endaseina (Hymenoptera, Ichneumonidae), II. *Endasys* Foerster 1868. Ann. Zool., Warszawa 45 (c. 60 pp., illustrated).

Lundbeck, W. (1896) 1897. Hymenoptera Groenlandica. Vidensk. Medd. Naturh. Foren. Kobenhavn, pp. 227-228.

Matthews, J.V., Jr. 1979. Tertiary and Quaternary environments: historical background for an analysis of the Canadian insect fauna. *In* H.V. Danks (ed.), Mem. Entomol. Soc. Canad. no. 108: 31-86.

————. 1980. Tertiary land bridges and their climate: backdrop for development of the present Canadian insect fauna. Canad. Entomol. 112(11): 1089-1103.

Mullier, P. 1979. Fecondité relative à l'âge des parasites de la tenthrède de Swaine (*Neodiprion swainei* Middleton) dans la province de Québec. M.S. thesis, Université Laval, Québec. 36 pp.

Nason, W.A. 1905. Parasitic Hymenoptera of Algonquin, Illinois, pt. 1. Entomol. News 16: 145-152.

Peirson, H.B., and R.W. Nash. 1940. Control work on European spruce sawfly in 1939. Bull. Maine For. Serv. no. 12: 16-17.

Perkins, J.F. 1962. On the type species of Foerster's genera (Hymenoptera: Ichneumonidae). Bull. Brit. Mus. (Nat. Hist.) 11: 383-483.

Price, P.W. 1970a. Characteristics permitting coexistence among parasitoids of a sawfly in Quebec. Ecology 51(3): 445-454.

————. 1970b. Trail odors: recognition by insects parasitic on cocoons. Science 170: 546-547.

————. 1971. Niche breadth and dominance of parasitic insects sharing the same host species. Ecology 52(4): 587-596.

————. 1972a. Parasitoids utilizing the same host: adaptive nature of differences in size and form. Ecology 53(1): 190-195.

————. 1972b. Activity patterns of parasitoids on the Swaine jack pine sawfly, *Neodiprion swainei* (Hymenoptera: Diprionidae), and parasitoid impact. Canad. Entomol. 104: 1003-1016.

————. 1973. Reproductive strategies in parasitoid wasps. Amer. Natur. 107: 684-693.

————. 1974a. Strategies for egg production. Evolution 28: 76-84.

——. 1974b. Evolutionary strategies of parasitic insects and mites. New York: Plenum Press. 224 pp.

——. 1975. Insect ecology. 2nd ed. New York: Wiley. 607 pp.

Provancher, L. 1874. Les ichneumonides de Québec avec description de plusieurs espèces nouvelles. Nat. Canad. 6: 173-177, 279-285.

——. 1875. Les ichneumonides de Québec . Nat. Canad. 7: 20-26, 74-84, 175-183.

——. 1877. Additions aux ichneumonides de Québec . Nat. Canad. 9: 5-16.

——. 1879. Faune canadienne; les insectes-Hymenoptères. Nat. Canad. 11: 65-76.

——. 1882. Faune canadienne; Hymenoptères; additions et corrections. Nat. Canad. 13: 321-336, 353-368.

——. 1883. Petite faune entomologique du Canada et particulièrement de la province de Québec, vol. 2, comprenant les Orthoptères, les Neuroptères, et les Hymenoptères. 830 pp.

——. 1886. Additions et corrections à la faune Hymenopterologique de la province de Québec; Fam. IV Ichneumonides, pp. 29-121.

——. 1887. Nos cantons de l'est. Nat. Canad. 16: 33-47.

Rau, D.E. 1976. Parasites and local distribution of cocoons of the yellowheaded spruce sawfly. M.S. Thesis, University of Minnesota, St. Paul. 44 pp.

Reeks, W.A. 1938. Native insect parasites and predators attacking *Diprion polytomum* (Hartig) in Canada. Entomol. Soc. Ontario Ann. Rept. 69: 27.

Riley, C.V., and L.O. Howard. 1890. Some of the bred parasitic Hymenoptera in the national collection. Insect Life 3(4): 153.

Roman, A. 1909. Ichneumoniden des Sarek Gebirges. Naturw. Untersuch. Sarekgebirges 4: 243.

——. 1913. Neubeschreibungen und Synonyme zur nördlichen Ichneumoniden fauna Schwedens. Entomol. Tidskr., pp. 112-132.

——. 1916. Ichneumonidae aus West-Grönland. Arkiv foer Zoologi 10(22): 4-5.

Say, T. 1836. Descriptions of North American Hymenoptera, and observations on some already described. Boston J. Nat. Hist. 1(2): 237-238.

Schedl, K.E. (1938) 1939. Die Populationsdynamik einiger kanadischer Blattwespen. Proc. 7th International Congr. Entomol. (1938) 3: 2052-2104.

Schmiedeknecht, O. 1890. Die Gattungen and Arten der Cryptinen revidirt und tabellarisch zusammengestellt. Entom. Nachr. 16: 1-150.

———. 1933. Opusc. Ichn., suppl.-bd. 2, fasc. 13-14: 18-54.

Short, J.R.T. 1959. A description and classification of the final instar larvae of the Ichneumonidae (Insecta, Hymenoptera). U.S. Natnl. Mus., Proc. 110 (3419): 391-511.

———. 1978. The final larval instars of the Ichneumonidae. Mem. Amer. Entomol. Inst. 25. 508 pp.

Slosson, A.T. 1897. Additional lists of insects taken in alpine region of Mt. Washington. Entomol. News 8: 237-240.

———. 1900. Additional list of insects taken in alpine region of Mt. Washington. Entomol. News 11: 319-323.

———. 1906. Additional list of insects taken in alpine region of Mt. Washington. Entomol. News 17: 323-326.

Smith, J.B. 1909. Insects of New Jersey. Ann. Rept N. J. State Mus. 888 pp.

Strand, E. (1926) 1928. Miscellanea nomenclatoria zoologica et palaeontologica. Arch. f. Naturgesch (A) 92(8): 30-75.

Strobl, G. (1900) 1901. Ichneumoniden Steiermarks (und der Nachbarländer). Mitt. Naturw. Ver. Steierm., Graz, 37: 132-157.

Swenk, M. 1911. A new sawfly enemy of the bull pine in Nebraska. Ann. Rept. Nebr. Agric. Exp. Sta. 24: 2-33.

Thomson, C.G. 1883. Foersoek till grupper im och beskrifning of Crypti. Opusc. Ent., fasc. 5: 455-530, fasc. 9: 843-936.

Thompson, L.C., and H.M. Kulman. 1980. Parasites of the yellowheaded spruce sawfly, *Pikonema alaskensis* (Hymenoptera: Tenthredinidae) in Maine and Nova Scotia. Canad. Entomol. 112(1): 25-29.

Townes, H.K. 1939. Corrections to the Gahan and Rohwer lectotypes of Provancher's Ichneumonidae (Hymenoptera). Canad. Entomol. 71: 91-95.

———. 1944. A catalogue and reclassification of the Nearctic Ichneumonidae (Hymenoptera), pt. 1; The subfamilies Ichneumoninae, Tryphoninae, Cryptinae, Phaeogeninae, and Lissonotinae. Mem. Amer. Entomol. Soc. 11(1): 1-477.

———. 1961. Types in Palearctic Museums. Entomol. Soc. Wash. Proc. 63: 103-113.

———. (1969 sic) 1970. Genera of Ichneumonidae, pt. 2. Gelinae. Mem. Amer. Entomol. Inst. 12. 537 pp.

——— and M.C. Townes. 1951. Family Ichneumonidae. *In* C.F.W. Muesebeck, K.V. Krombein, et al., Hymenoptera of America north of Mexico, synoptic catalog, USDA Monograph no. 2. pp. 184-409.

Tripp, H.A. 1960. *Spathimeigenia spinigera* Townsend (Diptera: Tachinidae), a parasite of *Neodiprion swainei* Middleton (Hymenoptera: Tenthredinidae). Canad. Entomol. 92: 347-359.

———. 1961. The biology of a hyperparasite, *Euceros frigidus* Cress. (Ichneumonidae) and description of the planidial stage. Canad. Entomol. 93: 40-58.

Underwood, G.R. 1960. Parasites of the red pine sawfly. Canad. Dept. Agric. For. Biol., Bi-monthly Prog. Rept. 16: 2.

Viereck, H.L. 1911. Descriptions of 6 new genera and 31 new species of ichneumon flies. U.S. Natnl. Mus. Proc. 40: 193.

———. 1914. Type species of the genera of ichneumon flies. U.S. Natn. Mus. Bull. 83. 186 pp.

———. (1916) 1917. Ichneumonidae. *In* Guide to the insects of Connecticut, pt. 3. The Hymenoptera, or wasp-like insects, of Connecticut, 22: 243-360.

Walkley, L.M. 1958. Family Ichneumonidae. *In* K.V. Krombein, Hymenoptera of America north of Mexico, synoptic catalog. USDA Monograph no. 2, suppl. 1: 36-62.

———. 1967. Family Ichneumonidae, pp. 60-212. *In* K.V. Krombein and B.D. Burks, Hymenoptera of America north of Mexico. Synoptic catalog. USDA Monograph no. 2, suppl. 2: 60-213.

Walley, G.S. 1931. *In* W.J. Brown, The entomological record, 1930. Entomol. Soc. Ontario. Ann. Rept. 61: 92.

Wray, D.L. 1967. Insects of North Carolina, 3rd suppl. N.C. Dept. Agric. 181 pp.

Figures

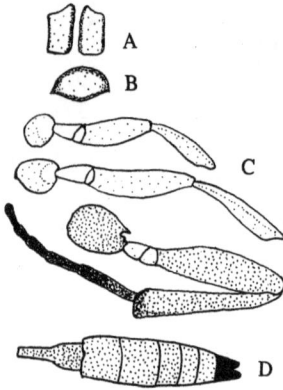

Fig 1

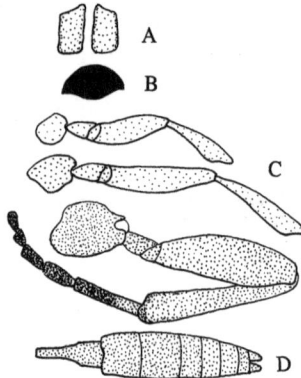

Fig. 2

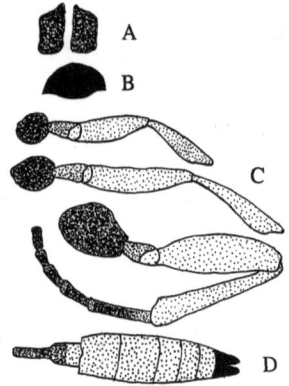

Fig. 3

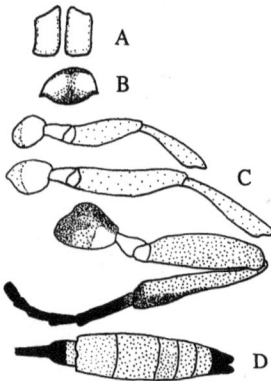

Fig. 4

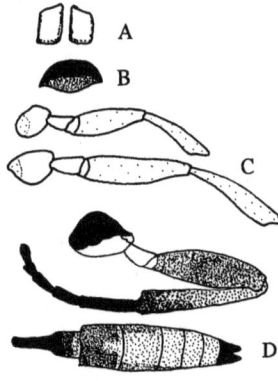

Fig. 5

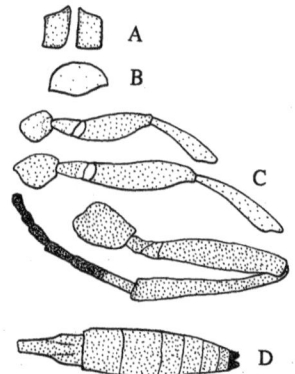

Fig. 6

Figs. 1-6. *Endasys* spp. Male (A) scape, (B) clypeus, (C) legs, and (D) abdomen.

1. *hesperus*, holotype. 4. *tricoloratus*, holotype.

2. *occipitis*, holotype. 5. *santacruzensis*, holotype.

3. *granulifacies*, holotype. 6. *leptotexanus*, holotype.

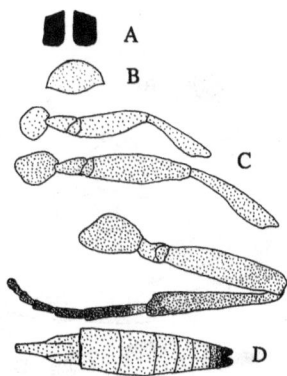

Fig. 7

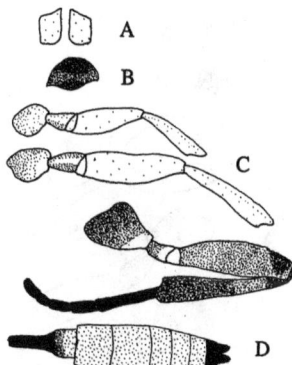

Fig. 8

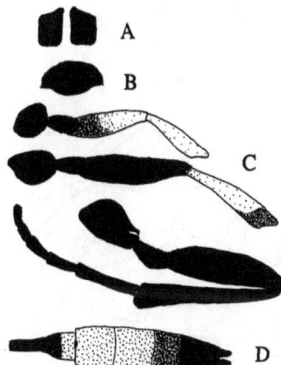

Fig. 9

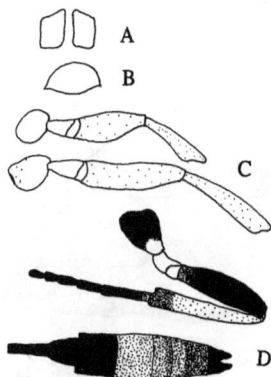

Fig. 10

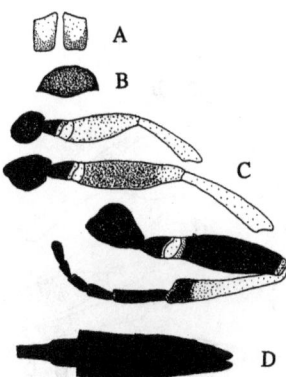

Fig. 11

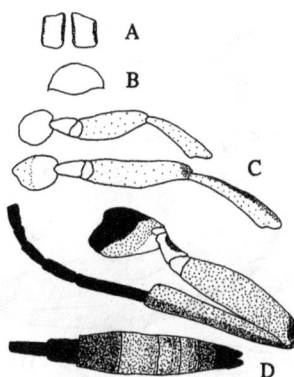

Fig. 12

Figs. 7-12. *Endasys* spp. Male (A) scape, (B) clypeus, (C) legs, and (D) abdomen.

7. *rubescens*, holotype.   10. *callidius*, holotype.

8. *tetratylus*, holotype.   11. *nigrans*, holotype.

9. *monticola*.   12. *gracilis*, holotype.

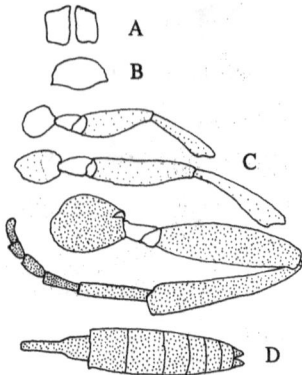

Fig. 13

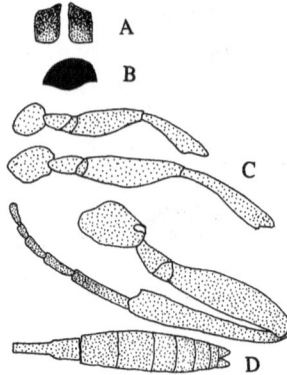

Fig. 14

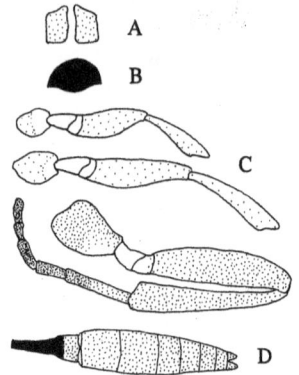

Fig. 15

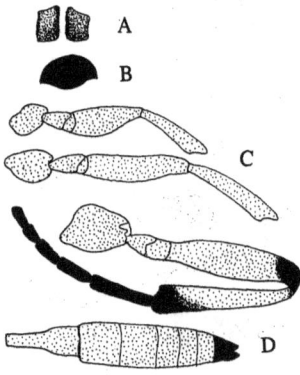

Fig. 16

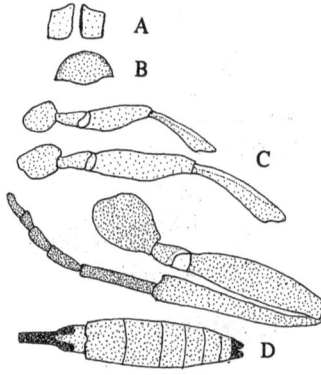

Fig. 17

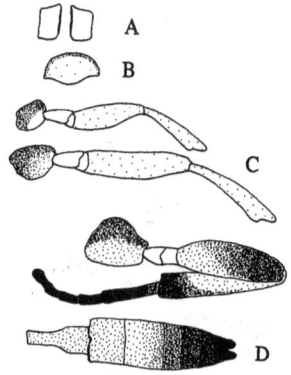

Fig. 18

Figs. 13-18. *Endasys* spp. Male (A) scape, (B) clypeus, (C) legs, and (D) abdomen.

13. *aurarius*, holotype.   16. *julianus*, holotype.

14. *aureolus*, holotype.   17. *auriger*, holotype.

15. *rugiceps*, holotype.   18. *brevicornis*, holotype.

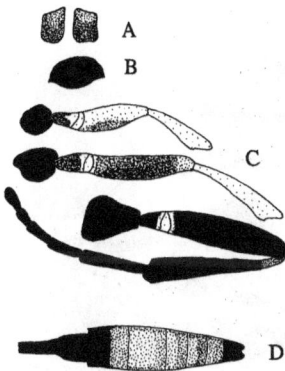

Fig. 19

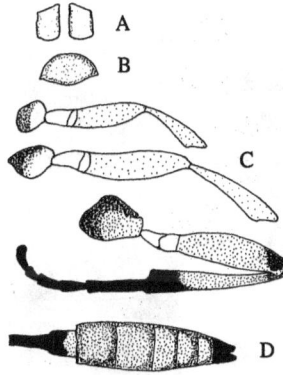

Fig. 20

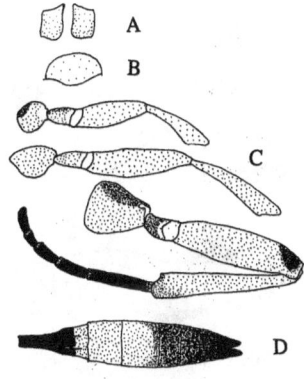

Fig. 21

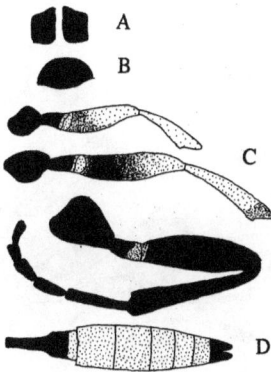

Fig. 22

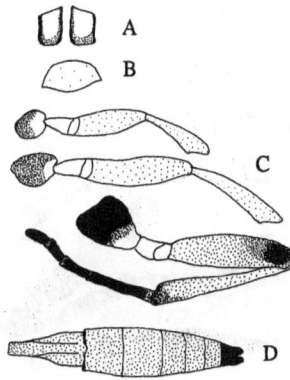

Fig. 23

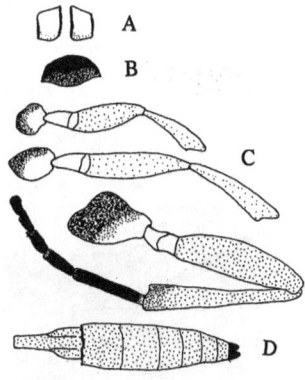

Fig. 24

Figs. 19-24. *Endasys* spp. (A) scape, (B) clypeus, (C) legs, and (D) abdomen.

19. *daschi*, holotype.   22. *serratus*, holotype.
20. *paludicola.*   23. *rugitexanus*, holotype.
21. *xanthopyrrhus*, holotype.   24. *texanus*, paratype.

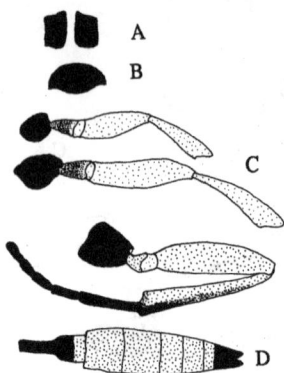

Fig. 25

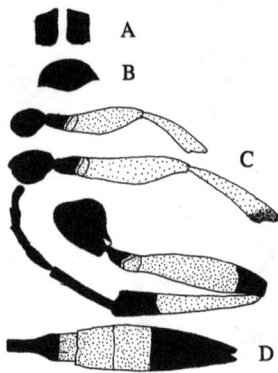

Fig. 26

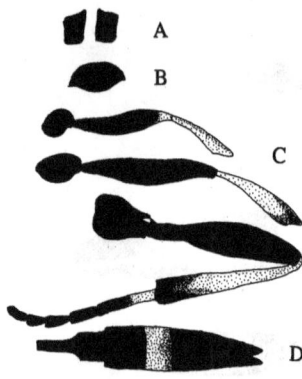

Fig. 27

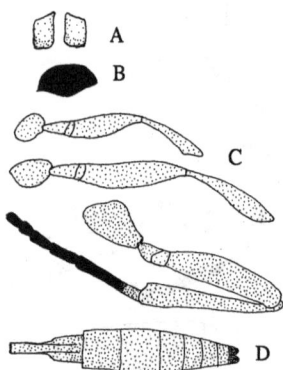

Fig. 28

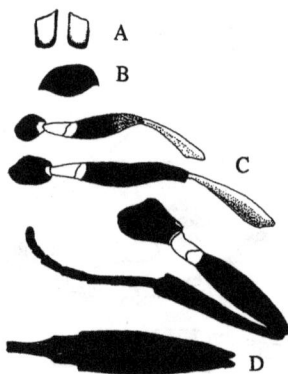

Fig. 29

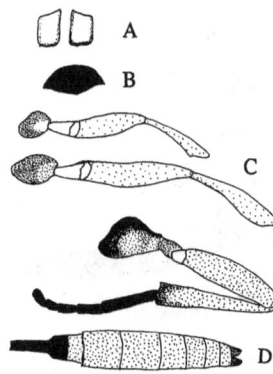

Fig. 30

Figs. 25-30. *Endasys* spp. Male (A) scape, (B) clypeus, (C) legs, and (D) abdomen.

25. *hexamerus*, holotype.   28. *oregonianus*, holotype.
26. *latissimus*, holotype.   29. *mucronatus*.
27. *melanurus*.   30. *concavus*, holotype.

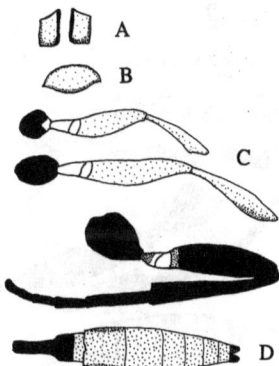

Fig. 31

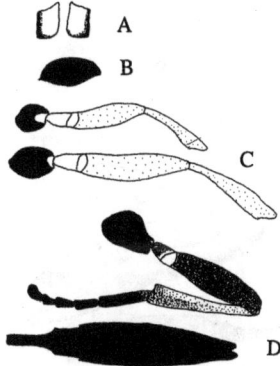

Fig. 32

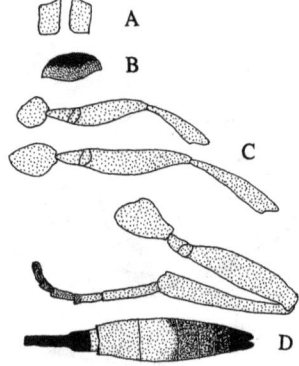

Fig. 33

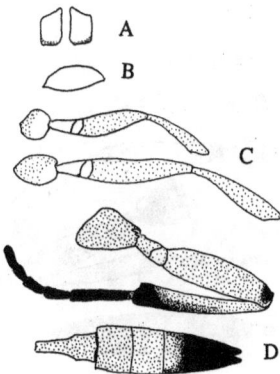

Fig. 34

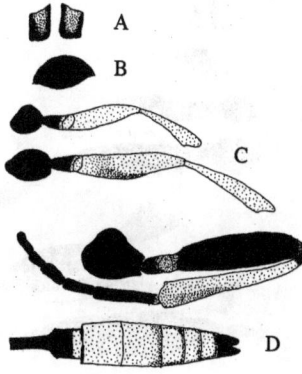

Fig. 35

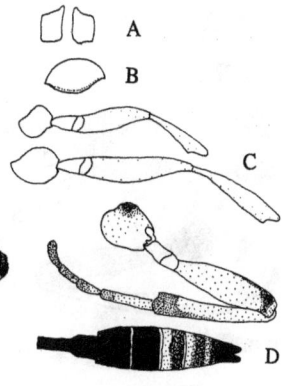

Fig. 36

Figs. 31-36. *Endasys* spp. (A) scape, (B) clypeus, (C) legs, and (D) abdomen.

31. *arizonae*, holotype.   34. *callistus*, holotype.

32. *melanogaster*, holotype.   35. *pentacrocus*, holotype.

33. *aurantifex*, holotype.   36. *elegantulus*, holotype.

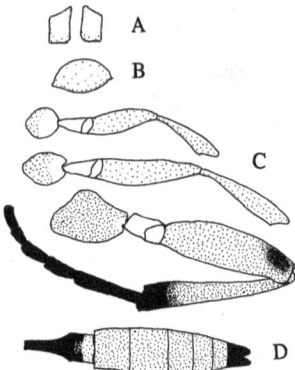

Fig. 37

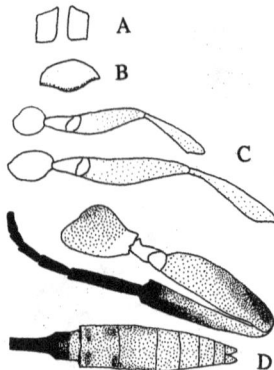

Fig. 38

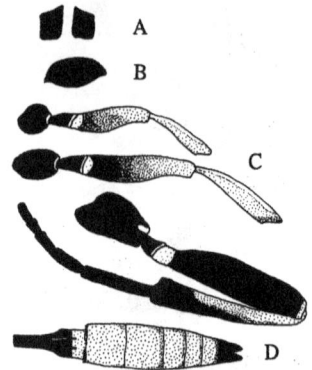

Fig. 39

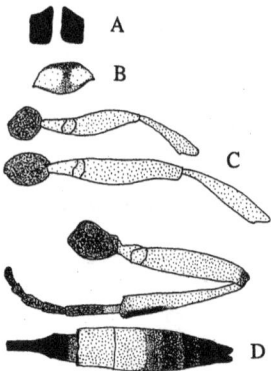

Fig. 40

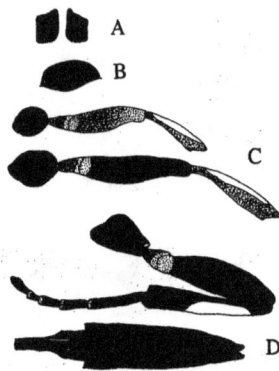

Fig. 41

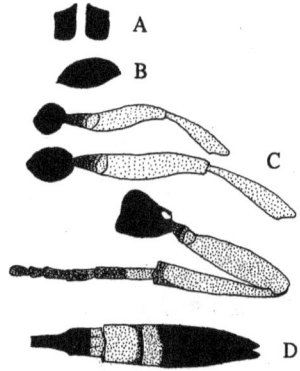

Fig. 42

Figs. 37-42. *Endasys* spp. Male (A) scape, (B) clypeus, (C) legs, and (D) abdomen.

37. *inflatus* homotype. 40. *xanthostomus*, holotype.

38. *tyloidiphorus*, holotype. 41. *maculatus*.

39. *taiganus*, holotype. 42. *coriaceus*, holotype.

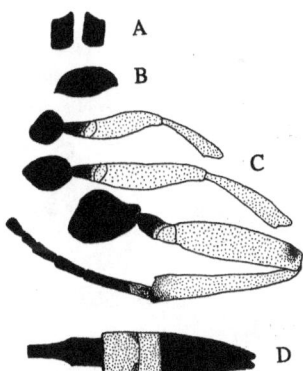

Fig. 43

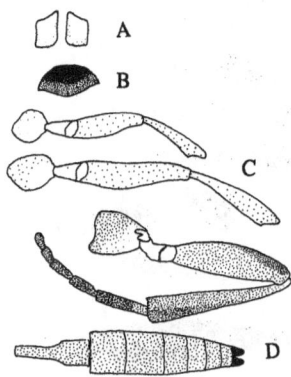

Fig. 44

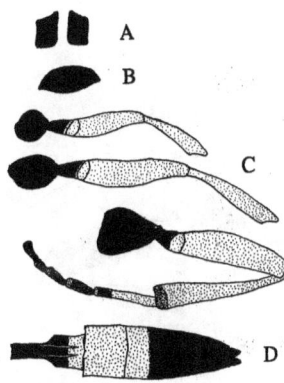

Fig. 45

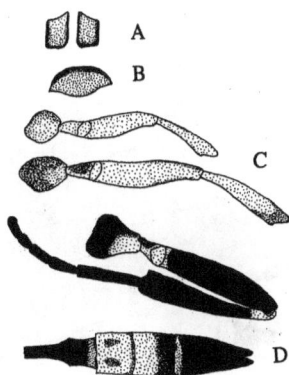

Fig. 46

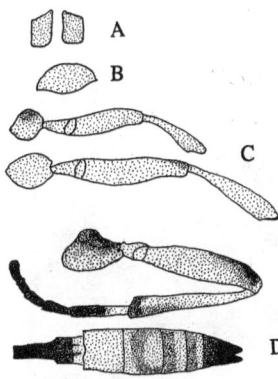

Fig. 47

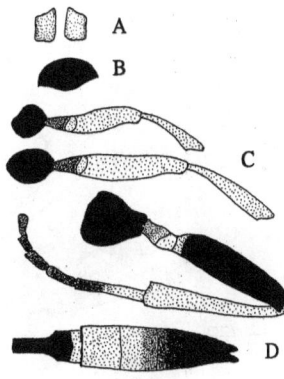

Fig. 48

Figs. 43-48. *Endasys* spp. (A) scape, (B) clypeus, (C) legs, and (D) abdomen.

43. *bicolor*.

44. *auriculiferus* homotype.

45. *declivis*, holotype.

46. *leopardus*, holotype.

47. *rhyssotexanus*, holotype.

48. *brachyceratus*, holotype.

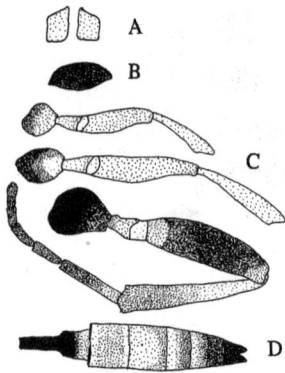

Fig. 49

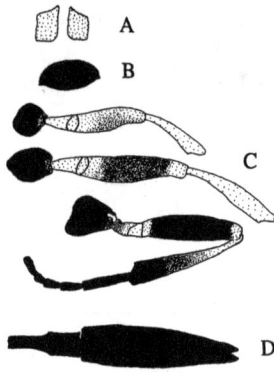

Fig. 50

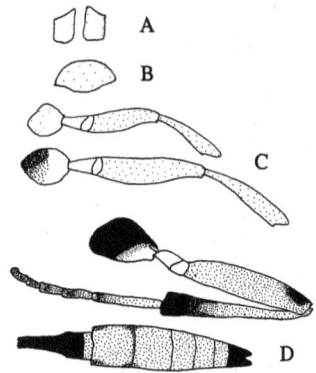

Fig. 51

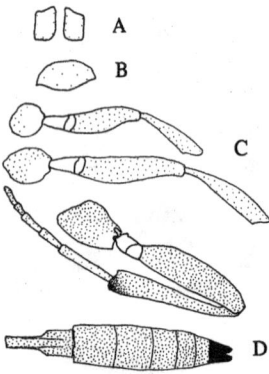

Fig. 52

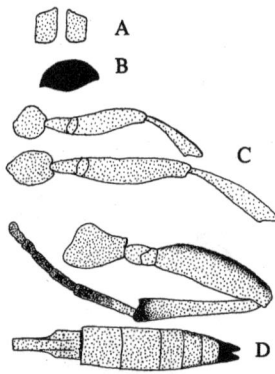

Fig. 53

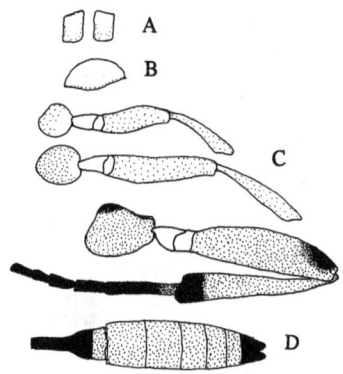

Fig. 54

Figs. 49-54. *Endasys* spp. Male (A) scape, (B) clypeus, (C) legs, and (D) abdomen.

49. *euryops*, holotype.   52. *subclavatus*.

50. *minutulus*.   53. *rugosus*, holotype.

51. *chrysoleptus*, holotype.   54. *aurigena*, holotype.

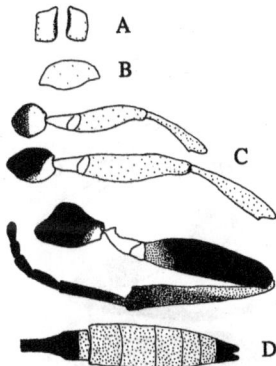

Fig. 55

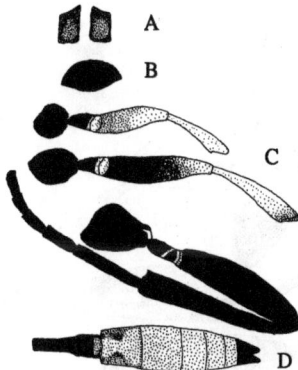

Fig. 56

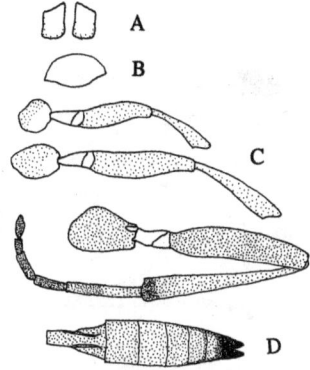

Fig. 57

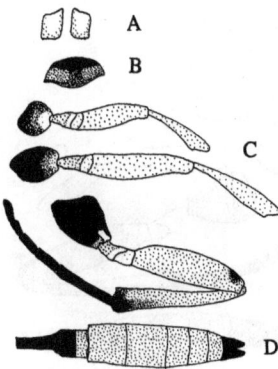

Fig. 58

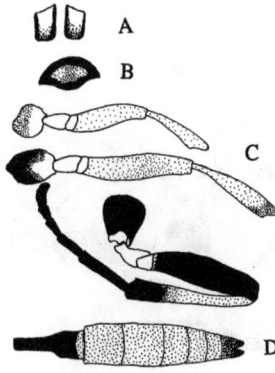

Fig. 59

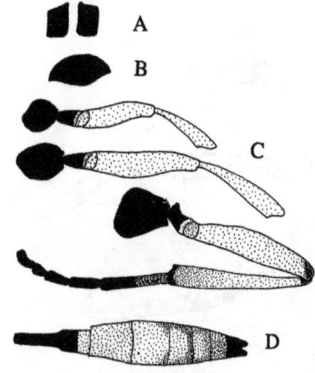

Fig. 60

Figs. 55-60. *Endasys* spp. (A) scape, (B) clypeus, (C) legs, and (D) abdomen.

55. *pubescens*, homotype.

56. *leucocnemis*, holotype.

57. *angularis*, holotype.

58. *chiricahuanus*, holotype.

59. *pinidiprionis*, holotype.

60. *leioleptus*, holotype.

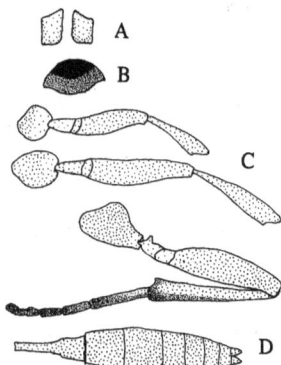

Fig. 61

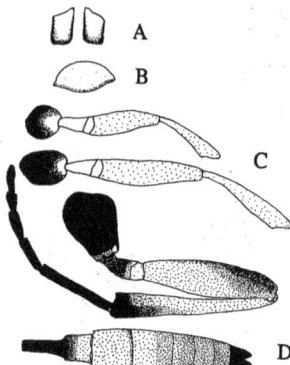

Fig. 62

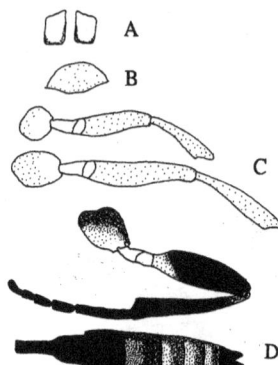

Fig. 63

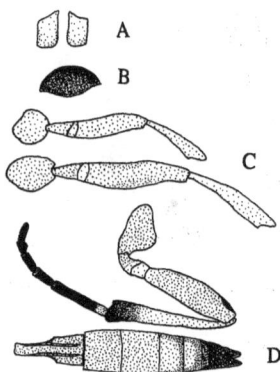

Fig. 64

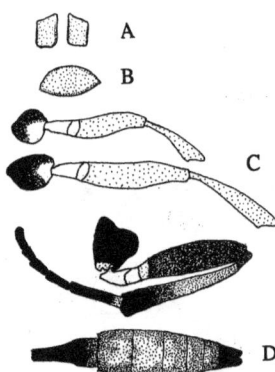

Fig. 65

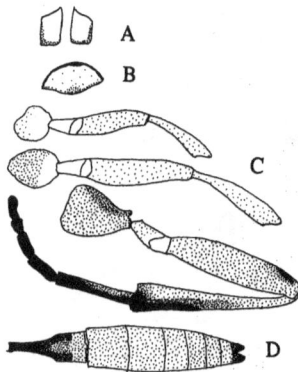

Fig. 66

Figs. 61-66. *Endasys* spp. Male (A) scape, (B) clypeus, (C) legs, and (D) abdomen.

61. *spicus*, holotype.
62. *albior*, holotype.
63. *praerotundiceps*, holotype.

64. *pseudocallistus*, holotype.
65. *nemati*, holotype.
66. *patulus*, homotype.

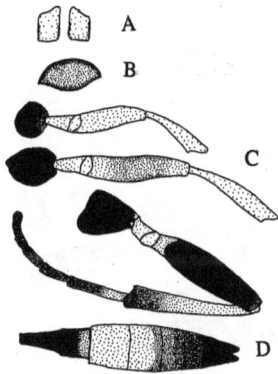

Fig. 67

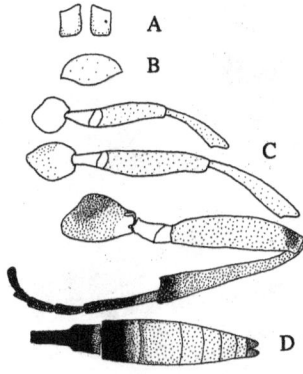

Fig. 68

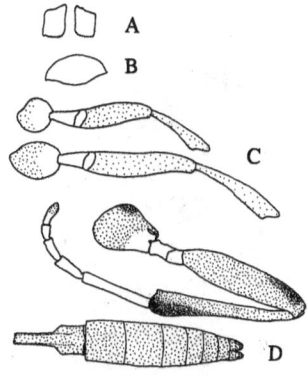

Fig. 69

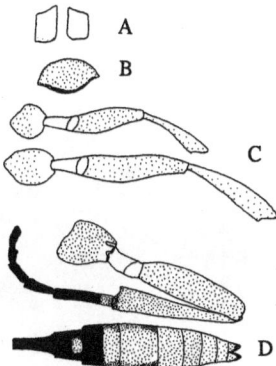

Fig. 70

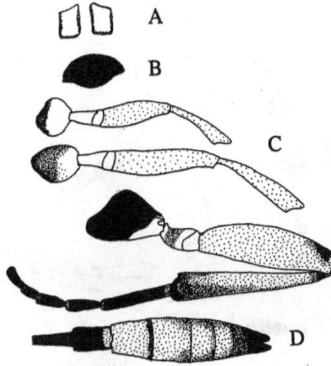

Fig. 71

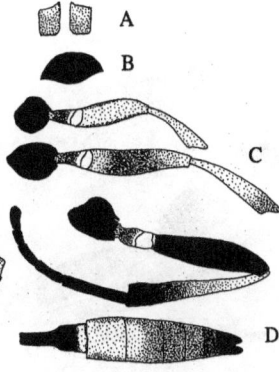

Fig. 72

Figs. 67-72. *Endasys* spp. (A) scape, (B) clypeus, (C) legs, and (D) abdomen.

67. *michiganensis*, holotype.   70. *arkansensis*, holotype.

68. *rotundiceps*.   71. *punctatior*, holotype.

69. *albitexanus*, holotype.   72. *obscurus*, holotype.

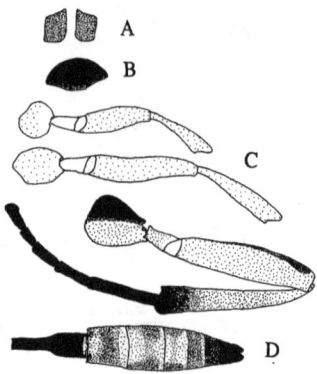

Fig. 73

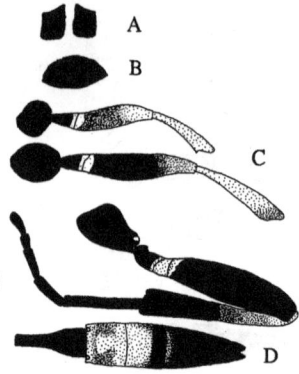

Fig. 74

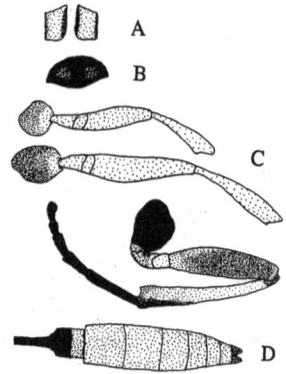

Fig. 75

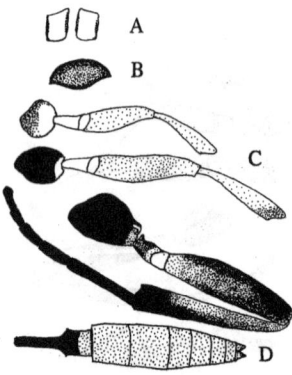

Fig. 76

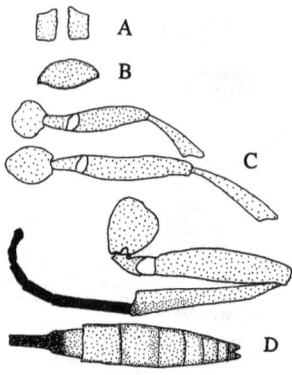

Fig. 77

Figs. 73-77. *Endasys* spp. Male (A) scape, (B) clypeus, (C) legs, and (D) abdomen.

73. *flavivittatus*, holotype.   76. *spinissimus*, holotype.

74. *bicolorescens*, holotype.   77. *flavissimus*, holotype.

75. *durangensis*, holotype.

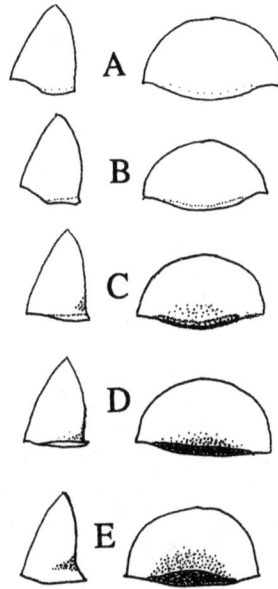

Fig. 78. *Endasys* spp. Clypeus (diagrammatic), lateral view (left column), anterior view (right column): (A) not upturned, (B) very weakly upturned, margin sharp, (C) weakly upturned, (D) distinctly upturned, (E) strongly upturned and impressed.

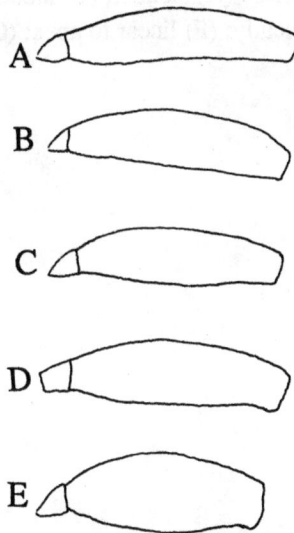

Fig. 79. *Endasys* spp. Femur III (diagrammatic), lateral view: (A) slender; (B) moderately slender; (C) moderately swollen (male) = moderately slender (female); (D) swollen (male) = moderately swollen (female); and (E) swollen (female).

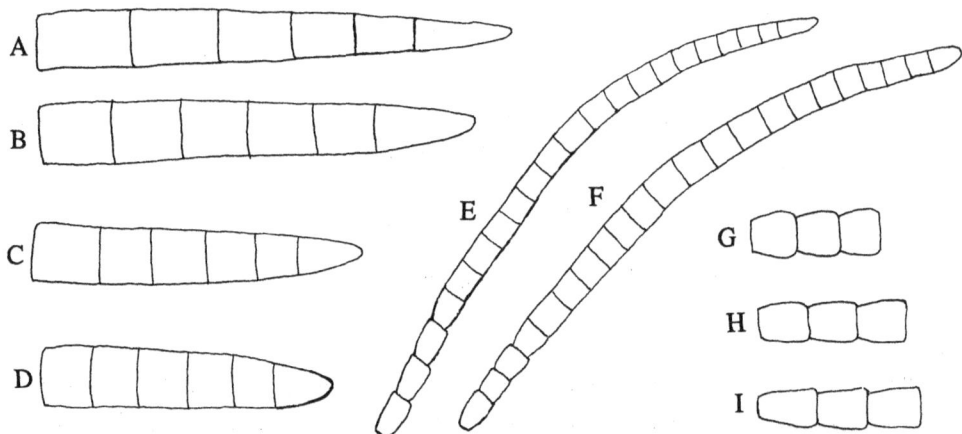

Fig. 80. *Endasys* spp. Flagellum (diagrammatic), (A-D) male apical segments, (E-F) female entire, (G-I) female basal 3 segments: (A) slender; (B) moderately slender; (C) moderately stout; (D) stout; (E) narrowed apically; (F) linear to apex; (G) moniliform; (H) short ; and (I) elongate.

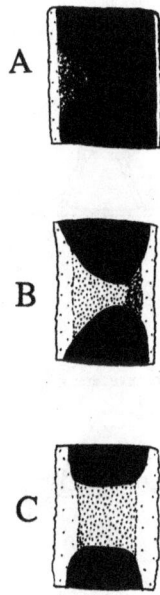

Fig. 81. *Endasys* spp. Male 5th abdominal sternum (diagrammatic): (A) complete, (B) apically complete, and (C) membranous.

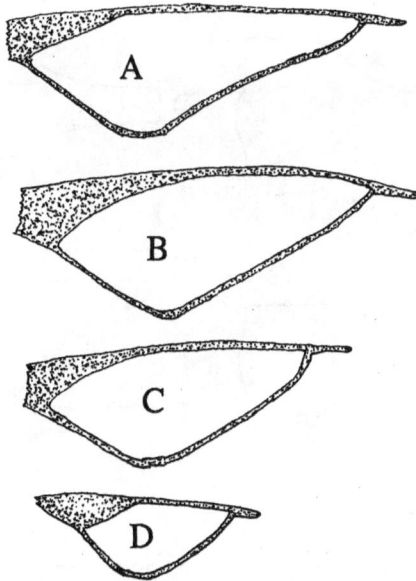

Fig. 82. *Endasys* spp. Radial cell: (A) slightly bowed, (B) straight, (C) curved apically, (D) short and curved.

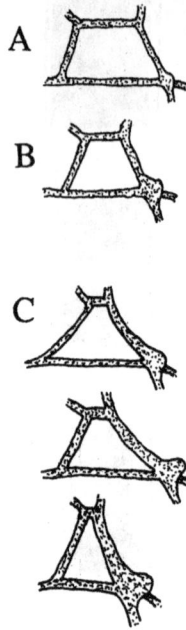

Fig. 83. *Endasys* spp. Area dentipara (diagrammatic) dorsal view: (A) elongate, (B) trapezoidal, (C) narrow to triangular.

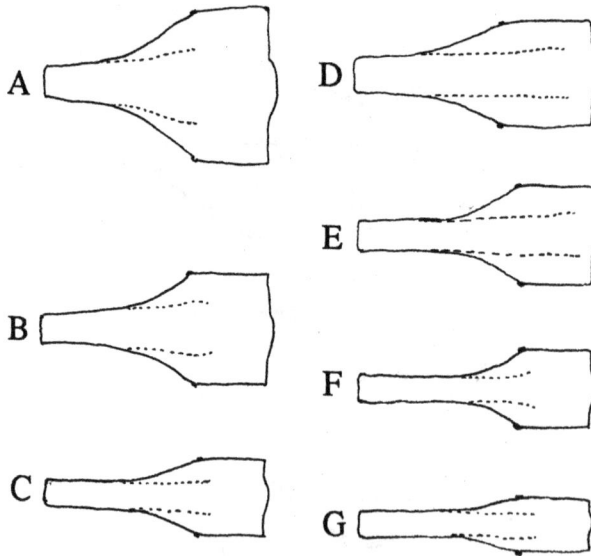

Fig. 84. *Endasys* spp. 1st abdominal tergum (diagrammatic) dorsal view, (A-C) female, (D-G) male: (A) stout, postpetiole wide; (B) moderately stout, postpetiole wider than long; (C) moderately slender, postpetiole square; (D) stout, postpetiole wide or square; (E) moderately stout, postpetiole wider than long; (F) moderately slender, postpetiole square; (G) slender, postpetiole narrow.

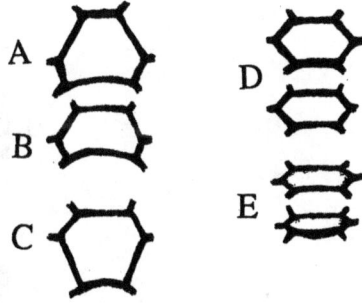

Fig. 85. *Endasys* spp. Areola dorsal view. (A) broadly hexagonal (female), (B) elongate (male), (C) narrow, (D) widely rectangular or (E) nearly linear.

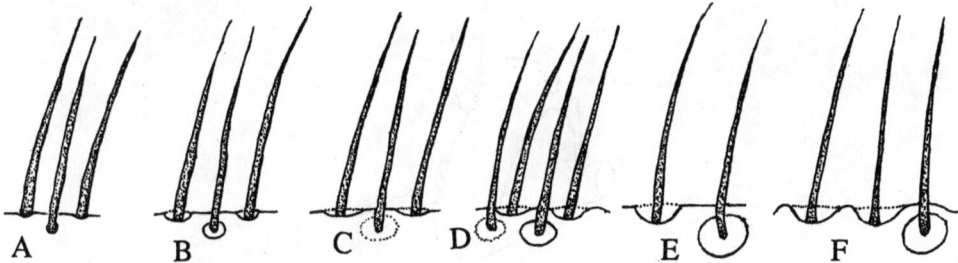

Fig. 86. *Endasys* spp. Setiferous punctation (diagrammatic), lateral and dorsal views. (A) smooth and shiny, no pits; (B) fine, pits sharp but small; (C) indistinct, pits shallow; (D) more or less distinct, some pits shallow, others deep; (E) distinct, pits sharp and deep; (F) coarse and rugulose, pits densely spaced and deep.

Fig. 87. *Endasys* spp. Tyloid on flagellomere (male): (A) long, (B) moderately long, (C) short, (D) faint.

Fig. 88. *Endasys* spp. Glumes on flagellomere (male): (A) dense, (B) moderately dense, (C) sparse.

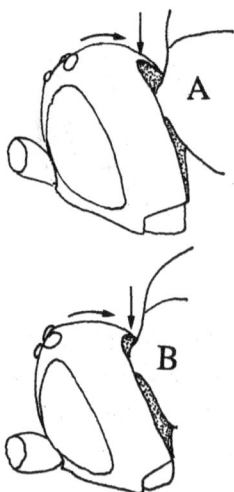

Fig. 89. *Endasys* spp. Occiput (arrow at occipital carina): (A) normal vertex curvature, (B) *occipitis*, vertex slightly flattened and lengthened.

Maps

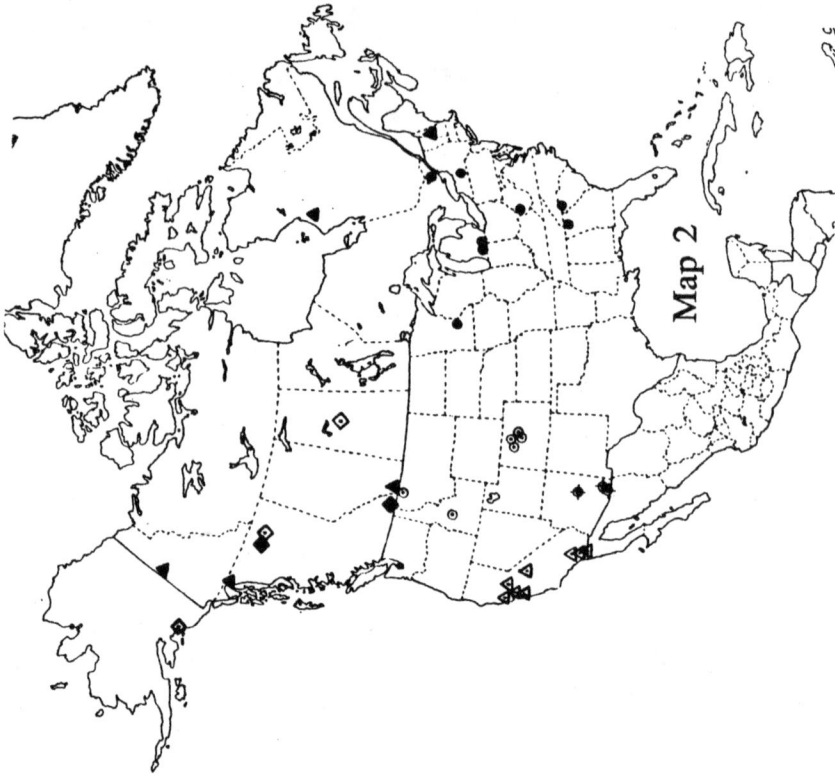

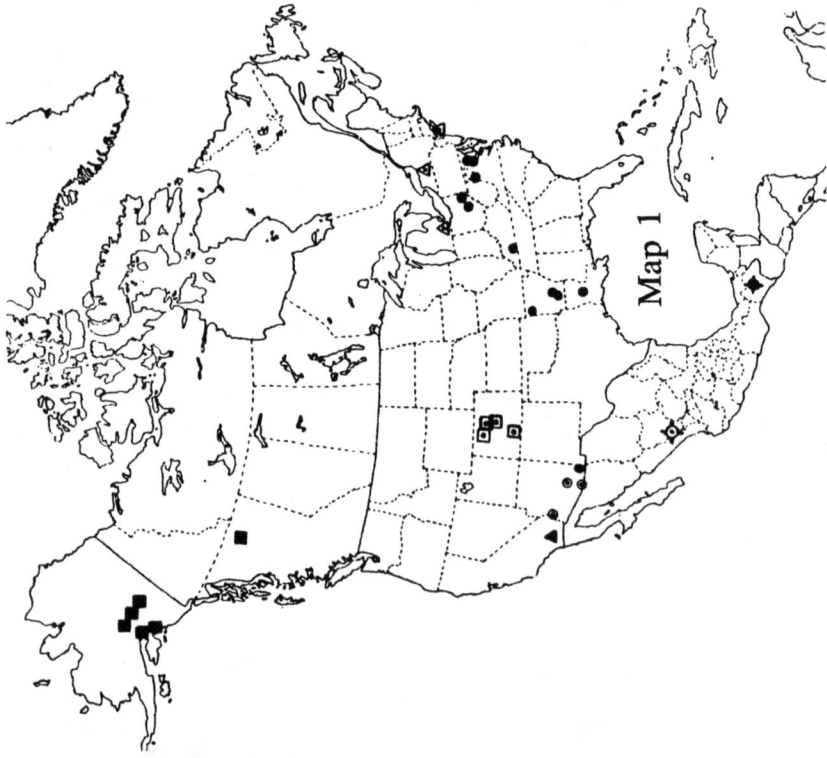

Map 1. *Endasys*: ⊡ *albior*, ▩ *minutulus*, △ *angularis*, ● *auriger*, ▲ *julianus*, ◆ *flavissimus*, ⊙ *arkansensis*, ✦ *spinissimus*.

Map 2. *Endasys*: ✦ *arizonae*, ● *albiexanus*, ◇ *coriaceus*, ▲ *bicolorescens*, ◆ *declivis*, ⊙ *daschi*, ◉ *flavivittatus*.

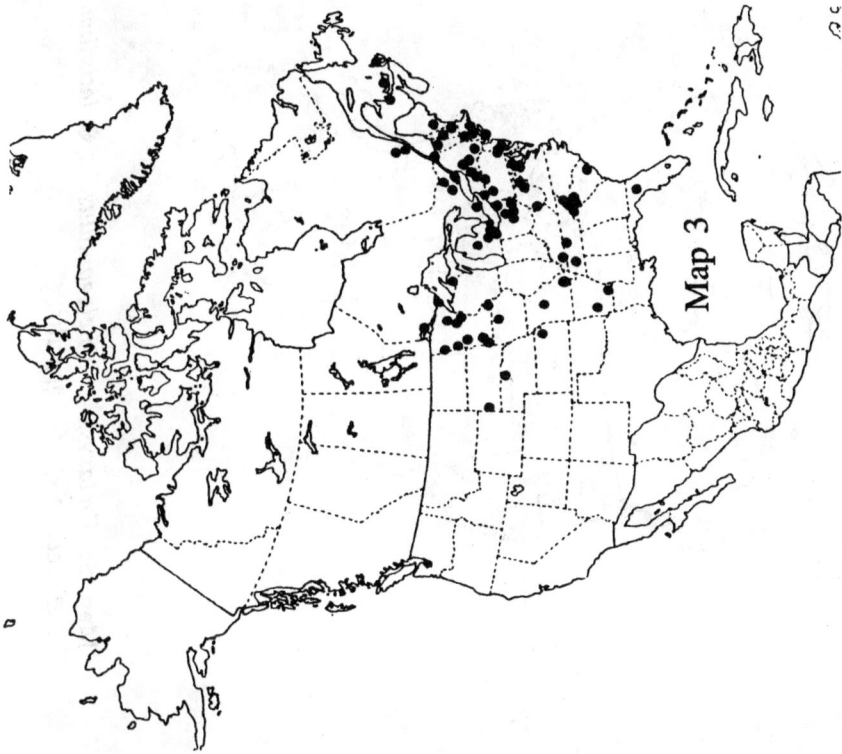

Map 3. *Endasys:* ● *auriculiferus.*

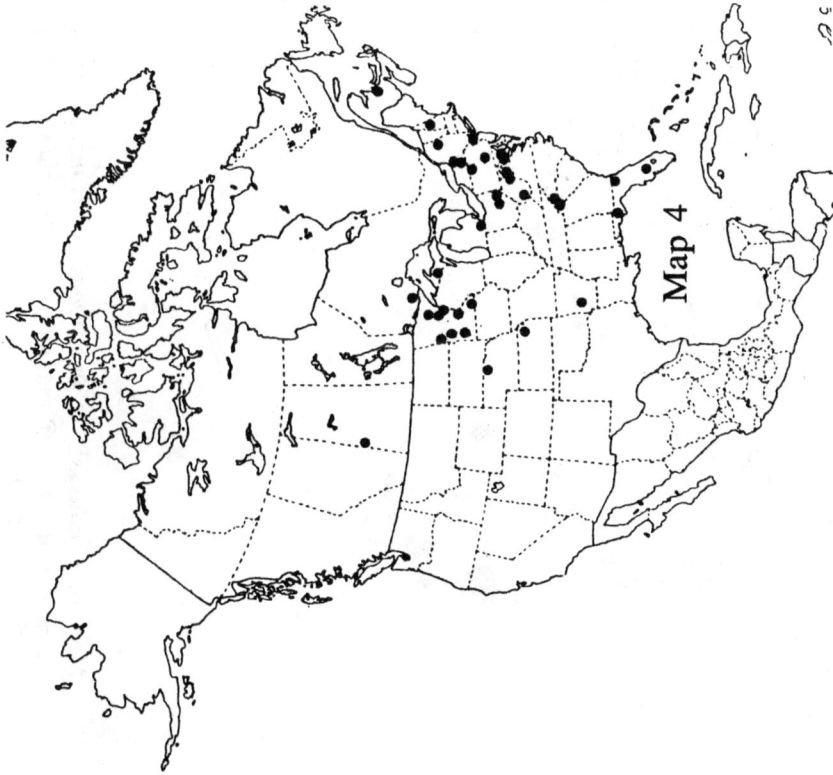

Map 4. *Endasys:* ● *aurigena.*

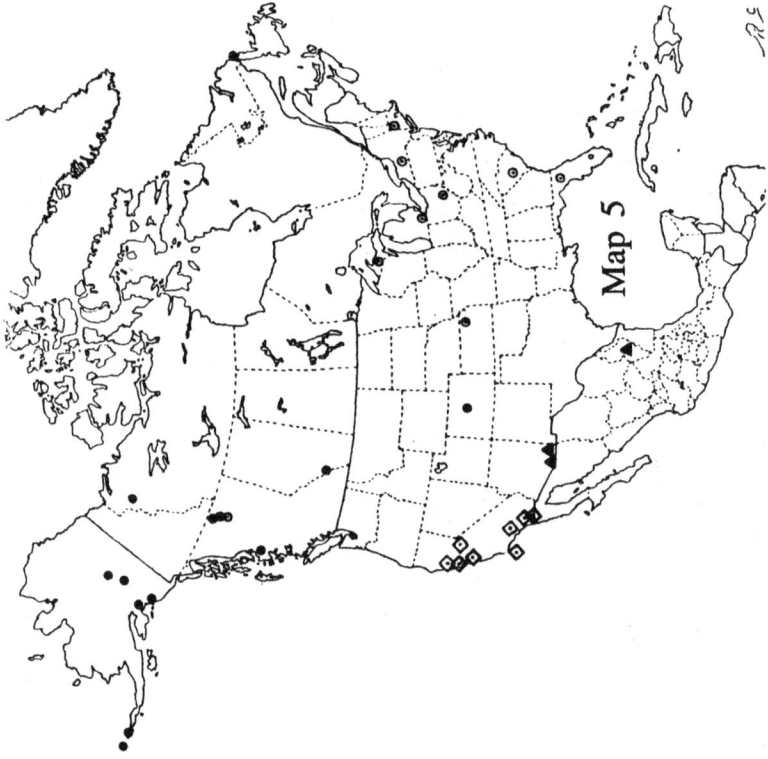

Map 5. *Endasys:* ⊙ *elegantulus*, ● *latissimus*,
▲ *occipitis*, ◈ *gracilis*.

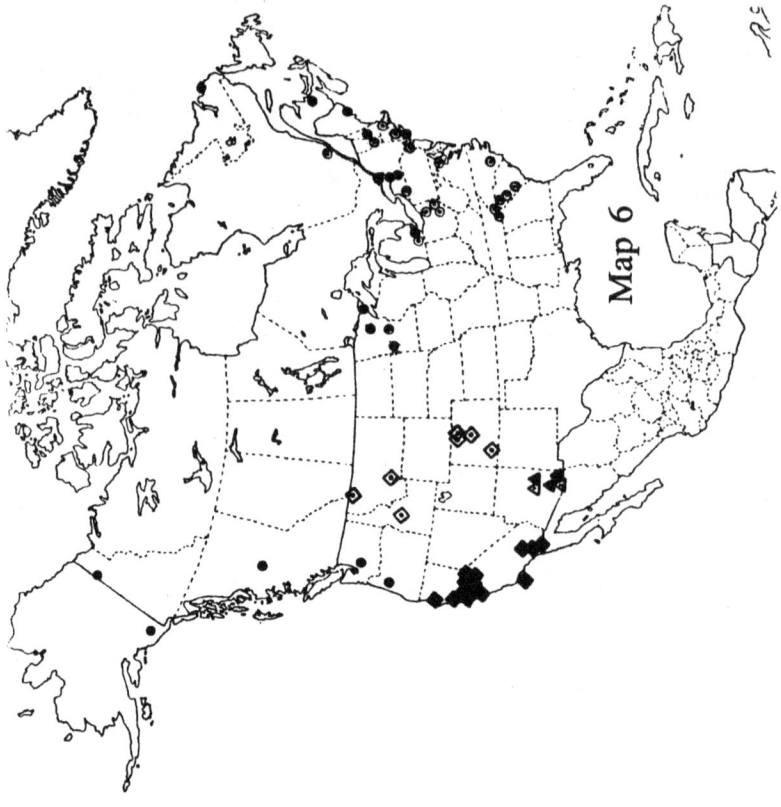

Map 6. *Endasys:* ● *euryops*, ⬠ *melanogaster*,
▲ *concavus*, ⊙ *maculatus*, ◈ *paludicola*,
◆ *punctatior*.

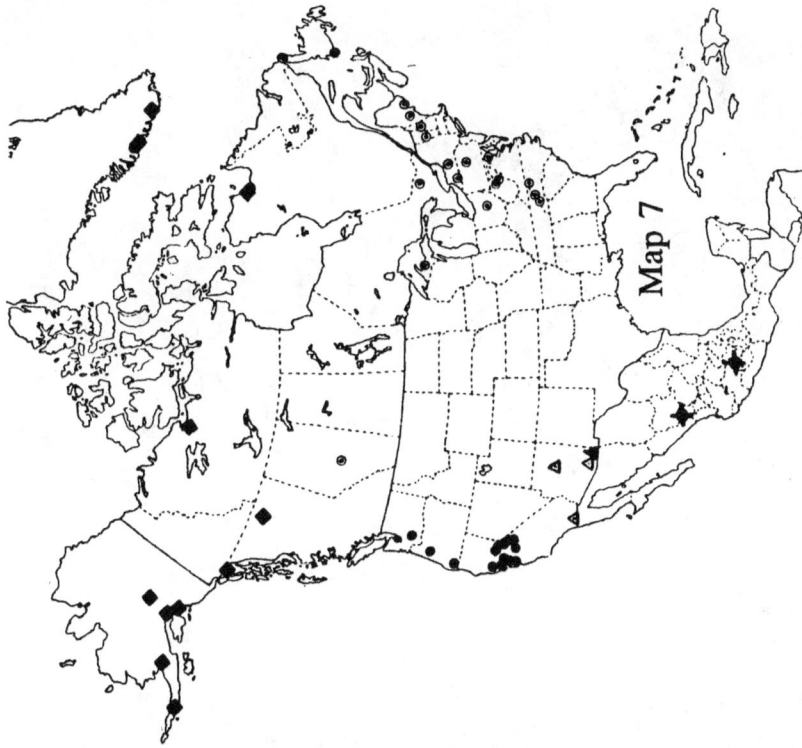

Map 7. *Endasys*: ● *granulifacies*, ▲ *chiricahuanus*, ◆ *bicolor*, ⬙ *callidus*, ◉ *chrysoleptus*, ✦ *durangensis*.

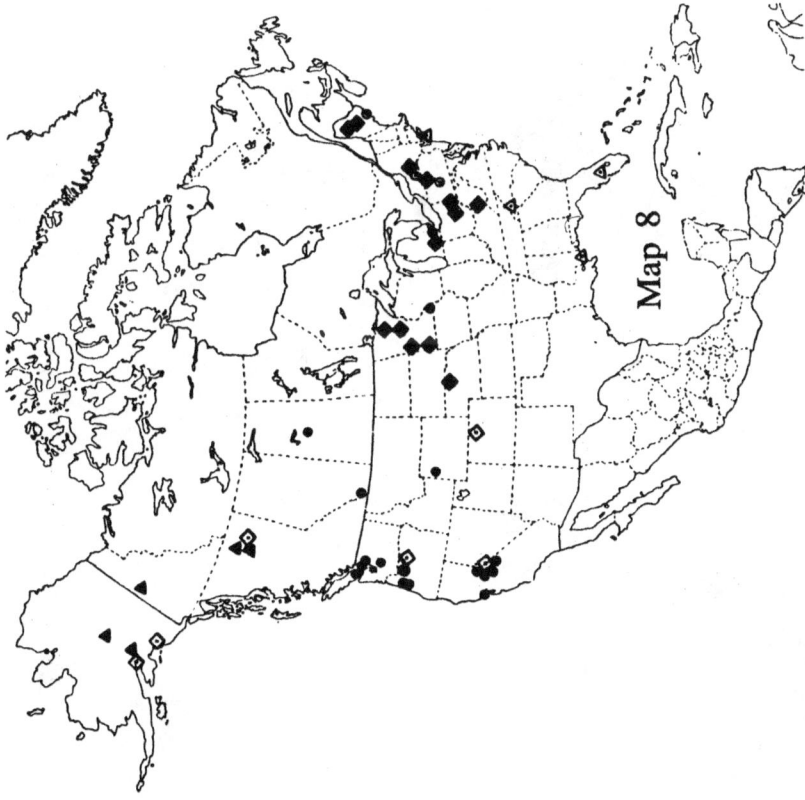

Map 8. *Endasys*: ● *hexamerus*, ◉ *pseudocallistus*, ◆ *callistus*, ◉ *rhyssotexanus*, ⬙ *spicus*, ◆ *brachyceratus*.

Map 10. *Endasys*: △ *leopardus*, ● *nigrans*,
● *michiganensis*, ◉ *rotundiceps*.
◆

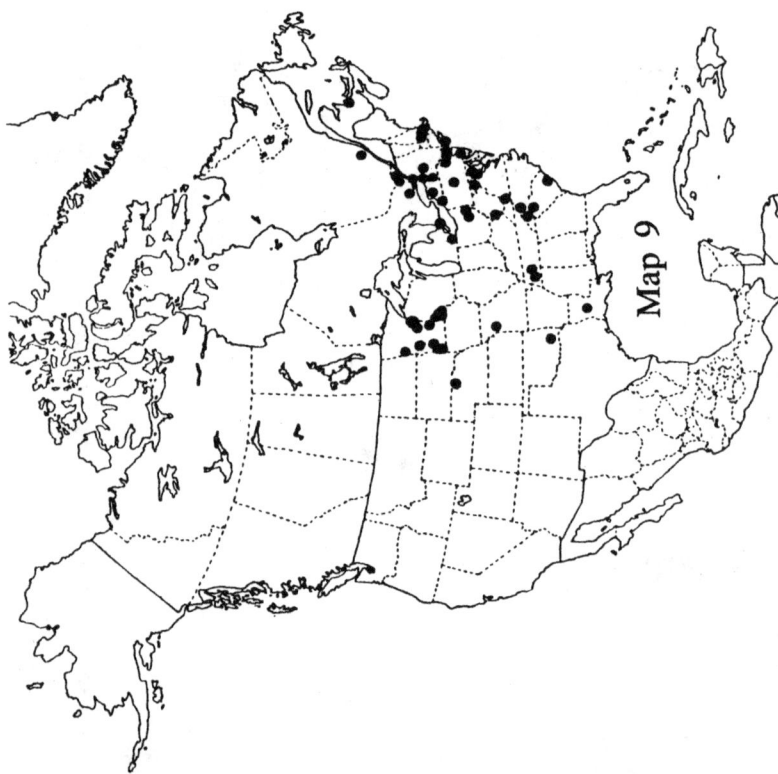

Map 9. *Endasys*: ● *inflatus*.

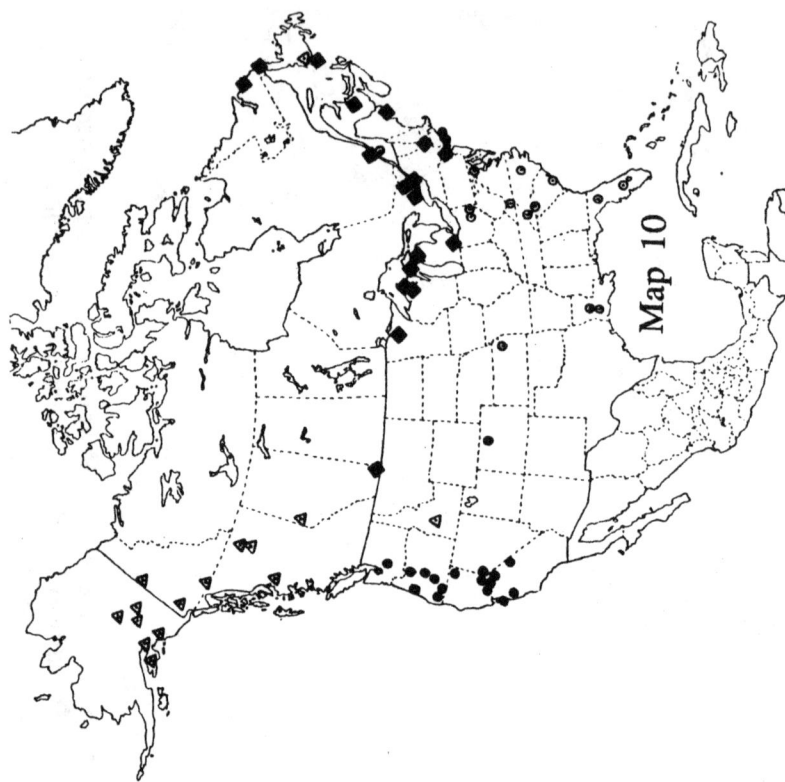

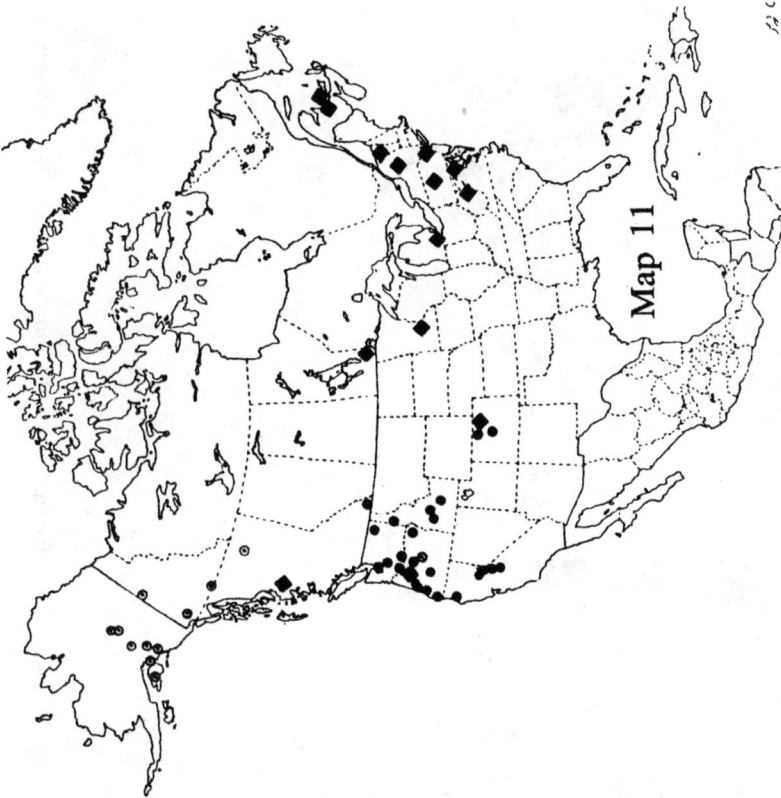

Map 11. *Endasys*: ● *leioleptus*, ⊙ *obscurus*, ◆ *rugosus*.

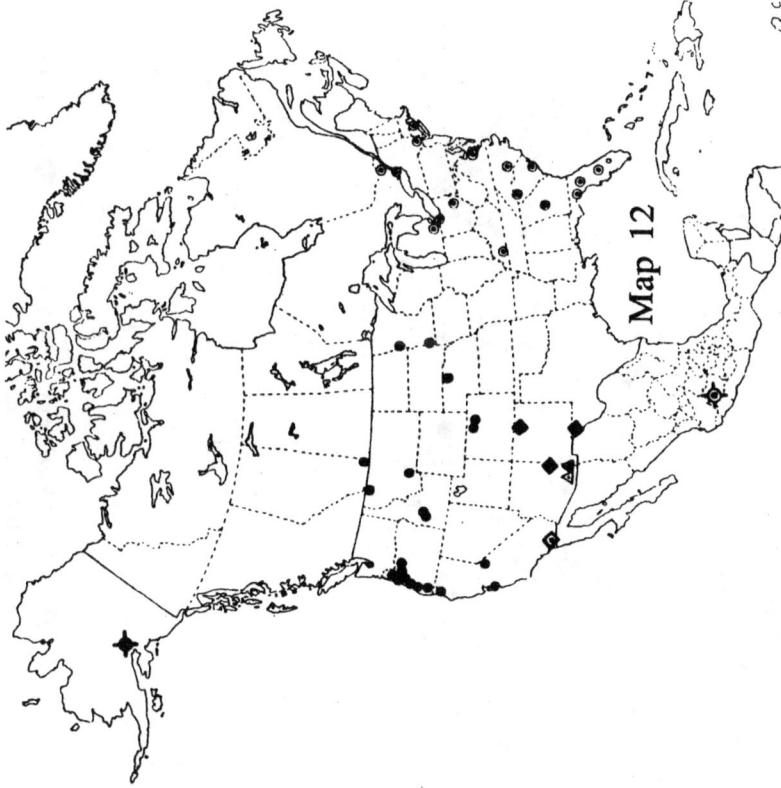

Map 12. *Endasys*: ✦ *leptotexanus*, △ *tetratylus*, ▲ *aureolus*, ● *aurantifex*, ⊙ *aurarius*, ◈ *tricoloratus*, ⊕ *pinidiprionis*, ◆ *leucocnemis*.

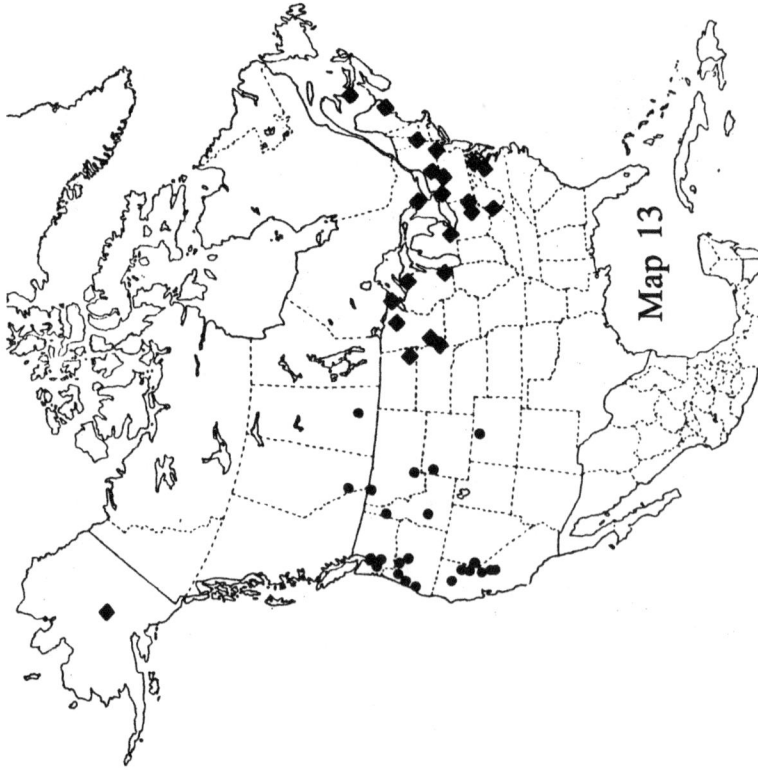

Map 13. *Endasys*: ● *monticola*, ◆ *rugitexanus*.

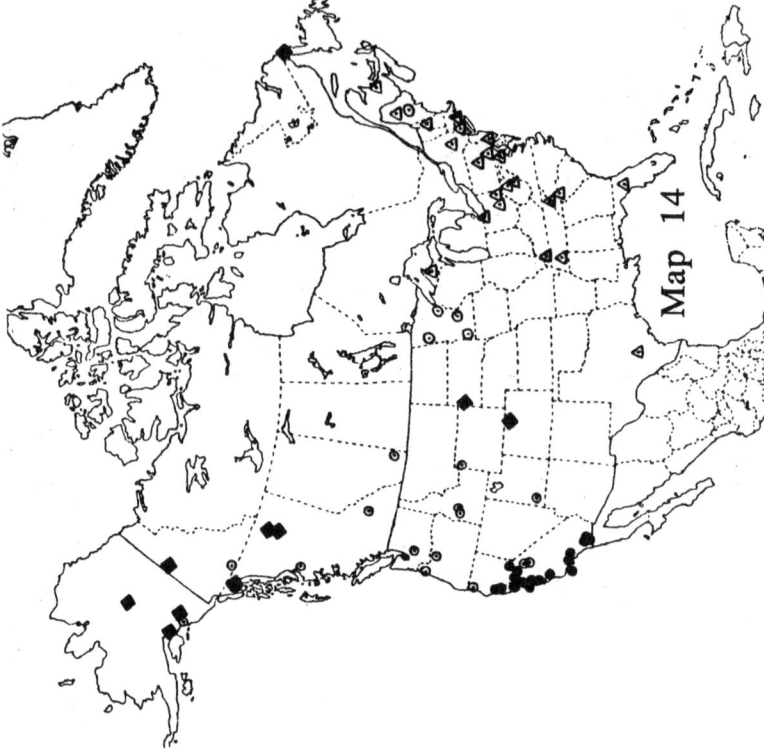

Map 14. *Endasys*: ⊙ *nemati*, ◆ *taiganus*,
● *santacruzensis*, △ *texanus*.

Map 16. *Endasys:* ● *pubescens.*

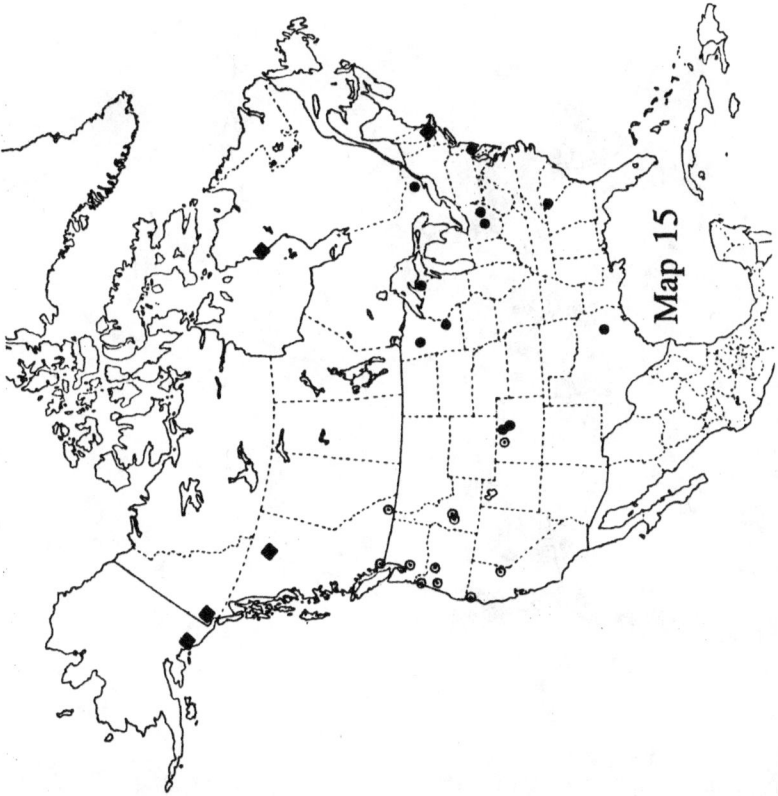

Map 15. *Endasys:* ⊙ *oregonianus,* ● *rugiceps,* ◆ *melanurus.*

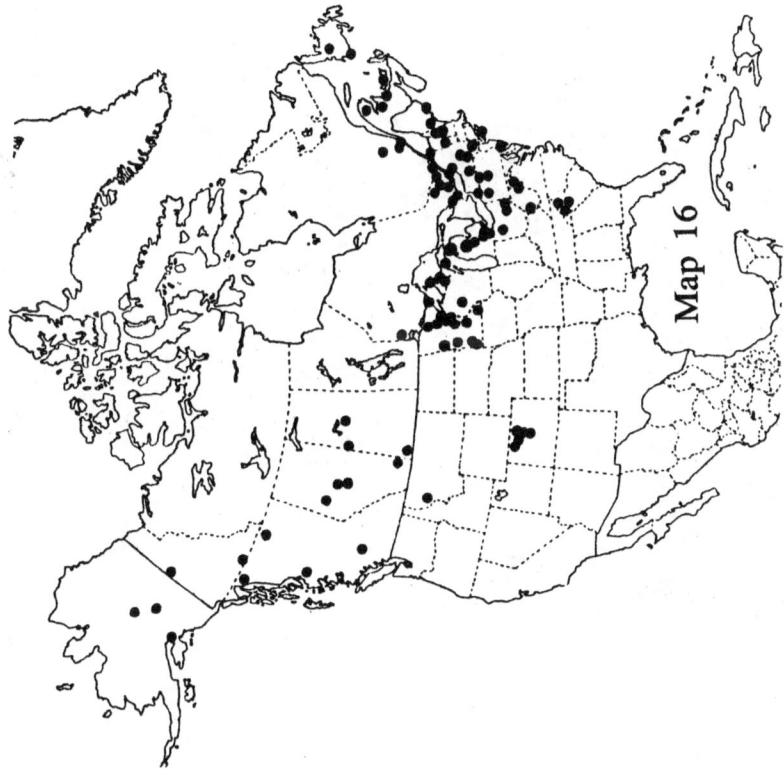

Map 17. *Endasys:* ●*praerotundiceps,* ◆*hesperus.*

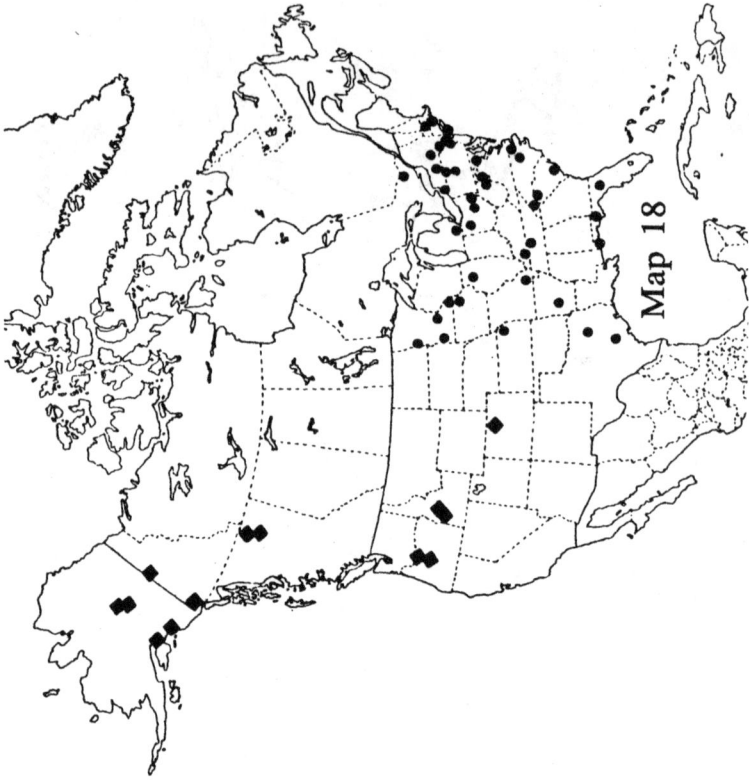

Map 18. *Endasys:* ◆*pentacrocus,* ●*tyloidiphorus.*

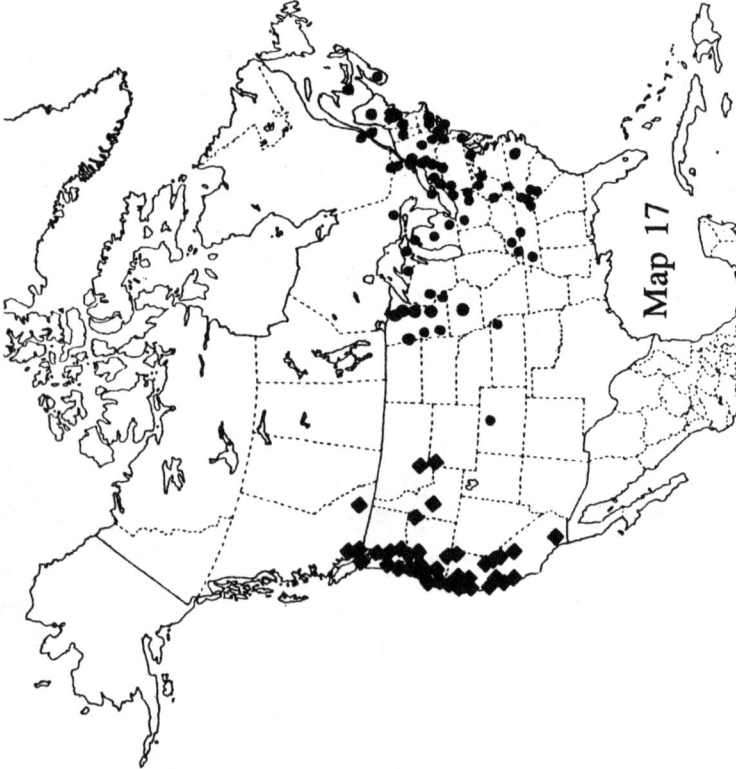

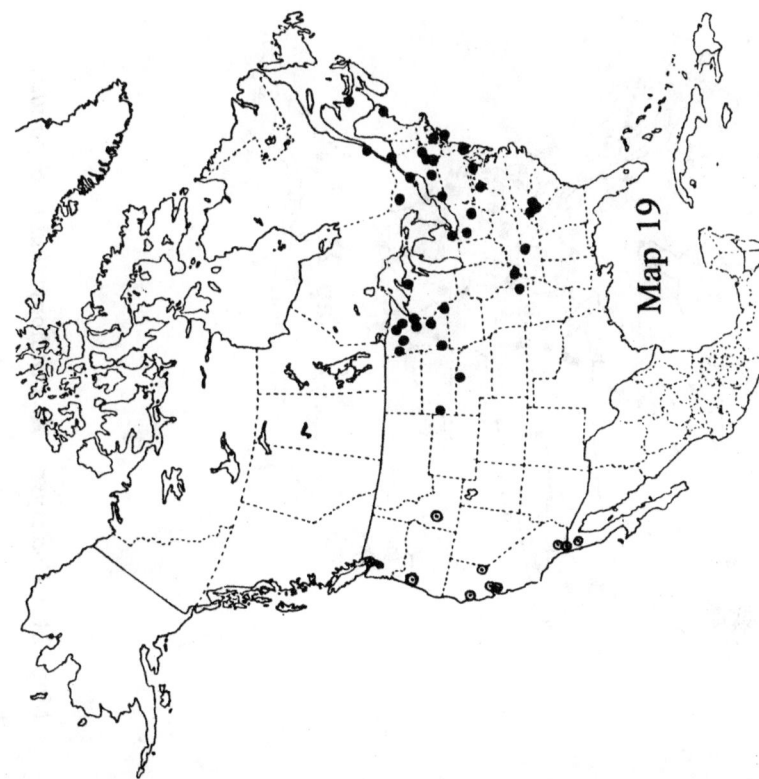

Map 19. *Endasys:* ⊙*rubescens*, ●*brevicornis.*

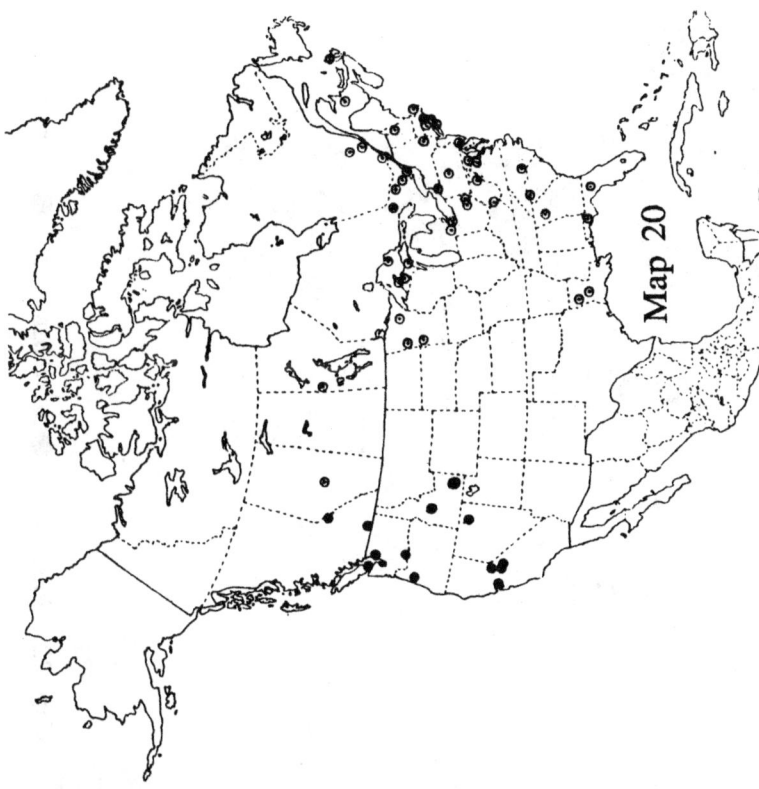

Map 20. *Endasys:* ●*serratus.* ⊙*mucronatus.*

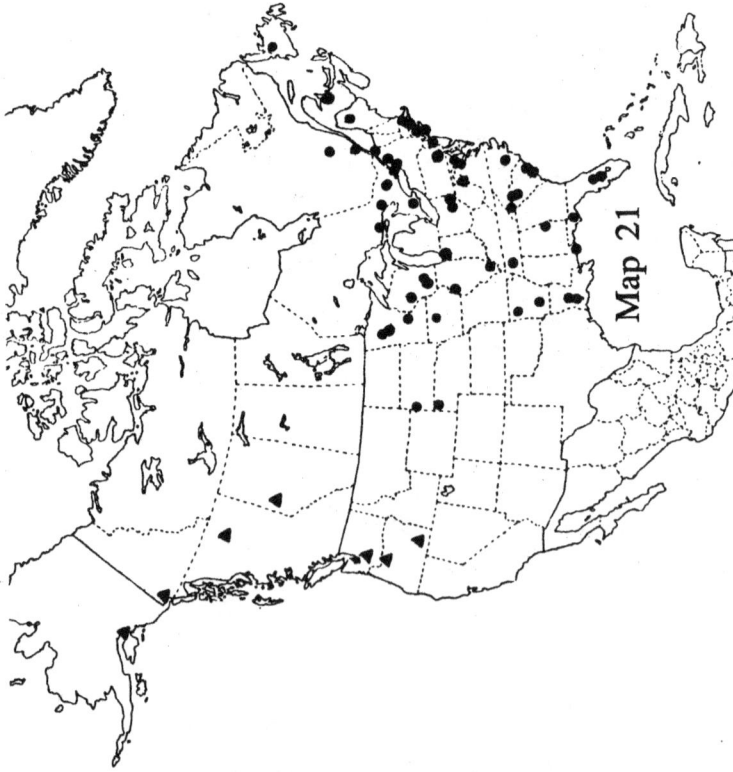

Map 21. *Endasys:* ▲ *xanthopyrrhus,* ● *patulus.*

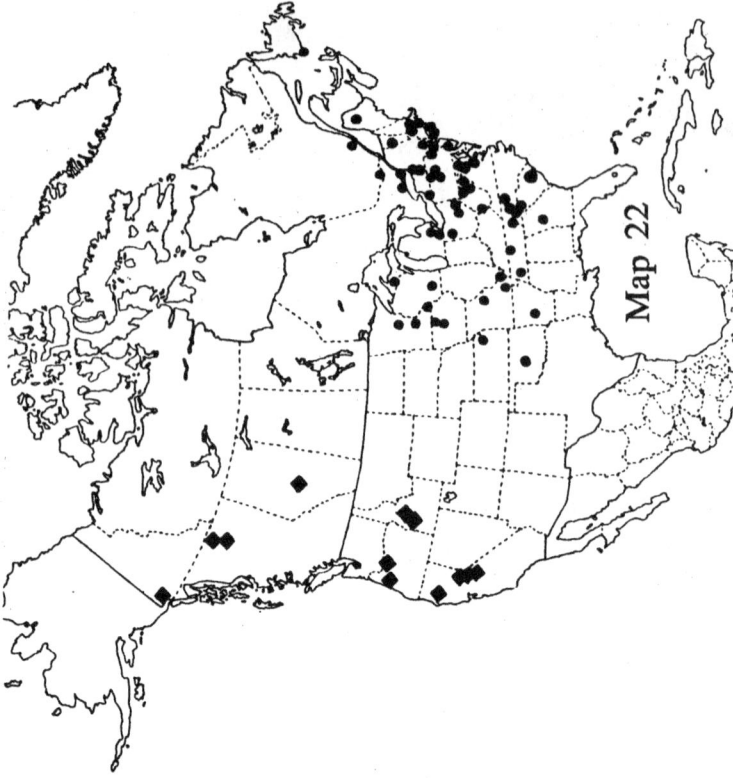

Map 22. *Endasys:* ◆ *xanthostomus,* ● *subclavatus.*]

www.ingramcontent.com/pod-product-compliance
Lightning Source LLC
Chambersburg PA
CBHW080419270326
41929CB00018B/3090